I0821854

GREAT DISCOVERIES IN SCIENCE

Electricity

Patrice Sherman

New York

Published in 2018 by Cavendish Square Publishing, LLC
243 5th Avenue, Suite 136, New York, NY 10016

First Edition

CPSIA Compliance Information: Batch #CS17CSQ

Library of Congress Cataloging-in-Publication Data

Names: Sherman, Patrice.
Title: Electricity / Patrice Sherman.
Description: New York : Cavendish Square Publishing, [2018] | Series: Great discoveries in science | Includes bibliographical references and index.
Identifiers: LCCN 2016053728 (print) | LCCN 2016055332 (ebook) | ISBN 9781502627728 (library bound) | ISBN 9781502627735 (E-book)
Subjects: LCSH: Electricity--Juvenile literature.
Classification: LCC QC527.2 .S534 2018 (print) | LCC QC527.2 (ebook) | DDC 537--dc23
LC record available at https://lccn.loc.gov/2016053728

Editorial Director: David McNamara
Editor: Caitlyn Miller
Copy Editor: Michele Suchomel-Casey
Associate Art Director: Amy Greenan
Designer: Lindsey Auten
Production Coordinator: Karol Szymczuk
Photo Research: J8 Media

The photographs in this book are used by permission and through the courtesy of: Cover Piotr Krzeslak /Shutterstock.com; p. 4 TebNad/iStock/Thinkstock; pp. 8, 70, 95 Bettmann/Getty Images; p. 13 Beeldbewerking /DigitalVision Vectors /GettyImages; p. 15 De Agostini Picture Library/Getty Images; p. 17 Duncan Walker /E+/Getty Images; pp. 20, 34, 59 Science & Society Picture Library/Getty Images; p. 24 PHAS/Universal Images Gro Getty Images; p. 28 Time Life Pictures/The LIFE Picture Collection/Getty Images; p. 38 Science Source /Getty Images; p. 42 Stanley K Patz/Photolibrary /Getty Images; p. 46 Georgios Kollidas/Shutterstock.com; p. 51 Hulton Archive/Getty Images; p. 54 NYPL/Scienc Source/Getty Images; p. 64 Library of Congress/digital version by Science Faction/Getty Images; p. 66 Tek Image Science Photo Library/Getty Images; p. 68 Rdj5150/iStock/Thinkstock; p. 73 Historical/Corbis Historical/Getty Ima p. 84 Photo 12/Universal Images Group/Getty Images; p. 88 Zephyr/*Science* Source; p. 91 MedicImage/Universal Images Group/Getty Images; p. 100 Alfred Pasieka/Science Photo Library/Getty Images.

Printed in the United States of America

Contents

Overhead electric lines are so ubiquitous most people hardly notice them until a discerning photographer turns them into art.

Introduction: Powering Our Lives

We call it juice. It fuels our power grids and fires up our computers. It comes alive with the press of a button or flick of a switch. It gives us heat and light and delivers sounds and sights from distant places to our phones and screens. It flows through the wires that run over our heads and beneath our streets. It even courses through every nerve in our bodies, enabling us to see, laugh, dance, think, run, read, smell, and dream. It can strike with deadly force and yet heal sickness. It is electricity—and it is everywhere. The story of the discovery of electricity stretches over twenty centuries, from ancient civilizations to the information age of today. It is a story that touches upon religion, philosophy, mechanics, engineering, physics, chemistry, biology, and mathematics. The people who created that story were scientists, inventors, detectives, dreamers, and occasionally showmen.

Electricity is so omnipresent in our lives that it's amazing to realize it was once considered a strange and unknowable force. To ancient people, electricity belonged to the gods. A flash of lightning was Thor's hammer or Apollo's spear. When the ancient Greeks rubbed a piece of amber against a cat's fur, they wondered at the static charge. Perhaps they had wakened a spirit inside, ready to jump out within a tiny spark.

The scientific study of electricity began in the eighteenth century. Benjamin Franklin helped establish the fact that lightning was really made of electricity. Pieter van Musschenbroek of the Netherlands learned how to store it in a jar. In Italy, Luigi Galvani made a frog's leg jump, and in 1800, Alessandro Volta put two and two together to make the world's first battery from a pile of zinc and copper coins. No one knew exactly what electricity was or how it worked, but the race to see what it could do was on.

Developments and discoveries came from surprising quarters. A London bookbinder's apprentice with scarcely any formal education realized he could generate electricity with a set of magnets and a copper coil. This insight made Michael Faraday the "father" of electromagnetic energy and one of the most famous scientists of all time. Meanwhile, in America, Joseph Henry, an Albany schoolteacher, sparked the idea for the telegraph when he unrolled a wire and made a bell ring.

During the last half of the nineteenth century, a group of remarkable inventors launched the great age of electricity. Telegraph wires sped messages across the continent and then across the world. Thomas Edison received US patent number 223,898 for an "electric lamp" that burned for a record 1,200 hours and would soon ensure "daylight" even at night. In 1893, George Westinghouse and Nikola Tesla lit up the Chicago World's Fair in a magnificent display of electric power that awed and delighted everyone who saw it.

Through all these events, a small group of research scientists worked quietly behind the scenes on the trail of the elusive **electron**. Too small to be seen, this infinitesimal bit of matter was the key that explained the great mystery that was electricity. Yet there remains much to be discovered. The same electricity that powers **generators** also powers our very brains. The same energy that surges through wires controls our own hearts. We, too, are made of electricity. In the twenty-first century, we stand at the edge of a new age of electricity—one that may merge minds with computers and open even more doors to the unsolved mysteries of electricity.

Benjamin Franklin's kite experiment is one of the most famous electrical experiments, but many historians doubt it actually happened.

CHAPTER 1

The Mystery of Electricity

One stormy June night in 1752, Benjamin Franklin went out to fly a kite fashioned from a silk handkerchief with a small metal hook affixed to the top. Franklin had attached a brass key to the kite's string within easy reach and brought with him a glass vessel lined with tin called a **Leyden jar**. He had enlisted the help of his twenty-one-year-old son, who huddled inside a nearby shed holding the end of the kite's string to keep it dry while Franklin stood outside and scanned the clouds for any sign of lightning. Though he did not see lightning strike, Franklin noticed some loose threads on the kite's string rise and wave about in the air as if pulled upward by a mysterious force. He concluded that the kite must have made contact with some hidden bolt of lightning in the dark clouds, and he touched the brass key with his knuckle. Sure enough, a spark flashed, instantly followed by a powerful jolt zipping through his arm. He held the Leyden jar up to the key and watched as it began to glow, filling up with what he called "electric fire." His hypothesis had been correct. Lightning was indeed made of electricity.

Franklin's adventure with his kite is one of the most famous scientific experiments in history. It has been depicted in countless paintings and drawings and is included in almost every biography written of him. Yet, some historians doubt that it ever happened—for a few reasons. First, a kite is not generally a good way to attract lightning. Kites may not fly high enough, and they bob and weave. If Franklin's kite was, as he described, no bigger than a large silk handkerchief, it was probably too small and light to support both the weight of the metal cross at its tip and the brass key dragging on its string near the ground. In short, it would never fly. And if it did, by some chance, manage to draw lightning, the resulting shock could have easily killed both Franklin and his son.

More puzzling, however, was Franklin's behavior after the supposed event. Franklin was a prolific writer who kept detailed records during his entire adult life. He published books, almanacs, journals, articles, and essays of all kinds. In 1751, his English friend Peter Collinson had even issued a short book based on letters Franklin had written to him concerning electricity, *Experiments and Observations on Electricity Made at Philadelphia in America by Benjamin Franklin.* Franklin was obviously interested in electricity and eager to communicate his ideas to the world. If he had discovered that lightning was a form of electricity, it seems likely he would have published an account of his experiment immediately or at least have written about it to a friend.

However, no report of Franklin's kite experiment appeared for nearly fifteen years. Even then, Franklin himself did not write it. In fact, the first public account of Franklin's experiment was published by the famous English chemist Joseph Priestly in 1767. Priestly claimed Franklin had told him of the experiment during a trip Franklin made to England in the early 1760s. Priestly hailed Franklin's realization that lightning and electricity were manifestations of the same force as a "capital" discovery, almost on a par with those of the great physicist

Isaac Newton. He did note, however, that, "This happened in June 1752, a month after the electricians in France had verified the same theory, but before he [Franklin] heard of anything they had done." According to Priestly, Franklin was not the first to make the discovery, but the second.

The Frenchmen to whom Priestly refers were Thomas-François Dalibard and Georges-Louis Leclerc. They had performed an experiment using a 40-foot (12.2 meter) metal pole to conduct electricity from lightning in May 1752. Dalibard had read Franklin's book, which included a detailed description of how one might use a kite to prove that lightning was made of electricity. The description does not say that Franklin had actually performed the experiment himself or intended to, only that it might be possible. Dalibard substituted a long metal pole for the kite and placed it in an empty wine bottle to ground it. During a thunderstorm on May 10, 1752, an assistant watched as lightning struck the pole and traveled down it into the empty bottle. Fortunately, lightning struck again and Dalibard and Leclerc were able to see the result. At one point, Dalibard touched the pole and burned his hand quite badly. A few days later, the men reported their findings to the French king and his court. From there, news of their discovery spread rapidly across Europe. May must have been a stormy month that year, for several others in both France and England had the opportunity to repeat the experiment within a few weeks. Today, historians acknowledge that Dalibard was the first to establish a connection between lightning and electricity.

In his memoirs, Franklin praises Dalibard for "drawing lightning from the clouds" and notes that he made the same experiment himself, "soon after with a kite at Philadelphia." He does not go into any detail regarding this event, though, and it remains unclear when or how it happened or if it actually did.

It remains a mystery why it is Franklin's name and not Dalibard's that we inevitably associate with the discovery that lighting is made of electricity. Perhaps the sheer drama of Franklin's story captures our imaginations. A fragile kite tossed about by fierce winds seems far more exciting than a mere metal pole. We imagine Franklin scanning the sky in the darkness, his kite bobbing into view and disappearing again behind the threatening clouds. Then, like a hero from an ancient legend he commands fire. At his touch sparks fly from the key on the kite string. Lightning was once assumed to be the weapon of the gods. Now Franklin has done what people have dreamed of doing for centuries. He has captured lightning and put it in a jar.

But Franklin did not know exactly what he had in that jar. Lightning may have been a form of electricity, but no one could say what electricity itself was. Electricity had fascinated and sometimes frightened people since ancient days. Over the centuries the study of electricity would touch upon religion, philosophy, science, and technology. Electricity raised questions about the nature of matter and the mysterious powers hidden in the universe. What was electricity made of, people wondered, and how did it travel? Most important of all, they wanted to know how its strange power and light could possibly be put to use.

The POWER of ATTRACTION: EARLY ENCOUNTERS with ELECTRICITY

Electricity has always been part of our world; humans noticed it long before they had a name for it. The first written references to electricity come from the ancient Egyptians. The Nile River contained a species of fish known as the electric catfish, which the Egyptians called the Angry Fish, Thunder Fish, or Protector of All Fish because of its ability to deliver

The ancient Vikings believed the god Thor caused thunder and lightning storms when he swung his hammer.

a hearty shock to any predator, including humans. Images of the fish appear on Egyptian murals and stone carvings, and it was sometimes associated with the pharaoh, indicating that the Egyptians may have believed that this god-like fish had the power to protect their god-ruler as well. Hieroglyphic texts imply that the Egyptians used small electric catfish to treat diseases such as epilepsy.

Other ancient civilizations were aware of electric fish as well. The Roman writer Pliny the Elder called it the "fish that shines by night" in his monumental work *Natural History,* completed around 79 CE. Like the Egyptians, the ancient Romans used electric fish to treat a number of conditions, from gout to migraine headaches. The sufferer was told to dip an arm into a pool containing a live electric catfish or ray and grab hold of it. The shock would supposedly cure the disease. Interestingly, electric shock therapy has been used in modern times to treat mental illness. It is extremely controversial but has helped some patients.

Magnets also fascinated ancient people. In prehistoric times, people had probably observed that some stones could attract other stones or small bits of iron and metal ores. They also would have noticed that two of these stones would either repel or attract one another depending on how they were positioned. Not surprisingly, people attributed magical powers to these special stones. They might place them on the body of an ill person in hopes of drawing out the sickness. Or the stones might be used in rituals to predict the future or called upon to help sages and shamans make the right decision.

As ancient societies became more sophisticated, people discovered what they believed was another kind of magnet. If a piece of amber was rubbed vigorously against the fur of a cat, it would attract feathers, straw, or bits of fiber. If placed close to someone's head, the amber could make hair stand on end. Amber is petrified tree resin; it is smooth and light and easily charged with **static electricity**. Static electricity occurs when two materials come into close contact with each other. Rubbing or friction can help generate static electricity but are not always necessary. When an object becomes charged it can attract or adhere to uncharged or neutral objects.

The Greek mathematician and philosopher Thales of Miletus wrote the earliest known description of static electricity around 600 BCE. He thought that by charging a

Greek philosopher and mathematician Thales of Miletus recorded some of the earliest observations about the nature of static electricity.

piece of amber he was turning it into a type of magnet—one that attracted straw rather than iron. Magnetism and electricity are not the same thing, though later scientists proved they can be related to one another through **electromagnetism**. It is easy to see, however, how they might have seemed to be manifestations of the same force.

Most ancient people believed that gods or spirits lived within magnets and amber, and this explained their powers. Thales had a slightly different theory. He believed that both magnets and pieces of charged amber were animated by a presence he called a soul.

He did not think this "soul" was the same as a god, though. It was more like a natural force, as much a part of the physical world as wind or water. Thales was not a scientist in the modern sense of the word, but he did believe in the pursuit of knowledge and recognized the importance of relying on one's own powers of observation to learn about nature.

MIRACLES, MAGIC, and SCIENCE: ELECTRICITY from the MIDDLE AGES to the RENAISSANCE

For hundreds of years after Thales, little, if anything, was written on either electricity or magnetism. Europeans appeared to have lost interest in investigating the natural world. During the Middle Ages, the works of the Greek and Roman philosophers were nearly forgotten. The Roman Catholic Church dominated Europe, and theology, or the study of religion, was considered the only subject worthy of study. If people studied the natural world at all, they did so only because it demonstrated the power of God. In the fourth century CE, Saint Augustine of Hippo, one of the most educated men of

MECHANICAL CONJURING TABLE, WITH APPARATUS. MR. NOVRA, REGENT STREET.

During the Middle Ages, performers often demonstrated magic tricks with static electricity at fairs and festivals.

his age, made some keen observations on the behavior of both magnets and static electricity. He even suggested that magnetism and static electricity were two separate, though related, forces. But he made it clear that he wrote about them only because he considered them miracles and proof of God's existence, not because they were important in their own right.

Magic was the dark side of religion. To the medieval mind, the world was filled with invisible and dangerous forces. Charged with static electricity, a piece of amber might very well be possessed by a demon or an imp. An electric shock could be an evil spirit jumping out of, or even worse, into the body of an innocent victim.

All kinds of superstitions surrounded the behavior of magnets, too, including the belief that the presence of fresh garlic would rob a magnet of its power. Sailors, for instance, would not eat garlic at sea for fear their breath would destroy the magnetic needle on the ship's compass. This conviction persisted until nearly 1600, when an English physician, William Gilbert, coated a number of magnets with garlic and proved once and for all that the pungent herb had absolutely no effect on magnetism. By then Europe had undergone enormous changes. The Middle Ages had given way to the Renaissance, an era of art, exploration, and learning. The study of the natural world became an important endeavor as rigid medieval theology was replaced by the more open philosophy of **humanism**.

Gilbert was mainly interested in magnets but he also investigated electricity. He was the first to use the term "electric" and noted that electricity traveled through some materials but not others. He believed that objects charged with electricity emitted a substance. He called it "effluvium." He didn't know what this substance was or how it moved, but he was convinced it was distinct from magnetism. (Today, we call it an electric current.) Gilbert's greatest contribution, however, was his use of the scientific method. In his introduction to his book on magnets, he wrote:

> In the discovery of secret things, and in the investigation of hidden causes, stronger reasons are obtained from sure experiments and demonstrated arguments than from probable conjectures and the opinions of philosophical speculators.

After Gilbert, experiments rather than religious belief or superstition would drive investigations into electricity. Once he had established that electricity could travel, the next questions were, how far? How fast? How strong?

ON with the SHOW: ELECTRICITY in the EIGHTEENTH CENTURY

In the 1720s, Stephen Gray, another Englishman, discovered that if he charged a glass tube with static electricity, the cork stopping the tube became charged, too. Intrigued, he inserted a stick with an ivory ball at the end from the cork and watched as the ball began to attract bits of thread and feathers. Next he replaced the stick with a silk thread and tied the ball to the end. He extended the thread 100, 200, 300 feet (30.5, 61, 91.4 m). Still the ball remained charged. In 1730, Gray and a friend performed the experiment again, conducting electricity from the tube along a single length of thread for a record 700 feet (213.3 m). His achievement won him a medal from Britain's Royal Society for Improving Natural Knowledge, the nation's first organization formed expressly to promote science.

Gray didn't stop with balls and strings. The human body, he knew, could also become charged with electricity. He recruited a boy from a nearby orphanage and proceeded

The Leyden jar was an early device for storing an electric charge. The jar was coated with metal.

to experiment on him. Gray suspended the child from the ceiling by ropes attached to his arms and legs. Then he brought the charged glass tube close to the boy's bare feet without quite touching them. Amazingly, the charge passed into the boy. Thin pieces of metal and other lightweight material scattered beneath him were drawn upward and adhered to his body. The boy survived this ordeal, and Gray's "Flying Boy" demonstrations became a regular event, attracting hundreds of wide-eyed, incredulous viewers.

After that, electric shows became all the rage in Europe. Nobody knew what electricity was but that didn't stop them from seeking it out. Suddenly, it seemed as if everyone wanted a shock. At one event, an electrically charged young woman set off sparks when she kissed a young man. "Fire flashed from her lips," a spectator wrote, leaving both parties with some very painful blisters. Nevertheless, the performances continued. Another standard stunt consisted of passing an electric charge through a group of people holding hands in a circle, making them all jump at once, much to the amusement of onlookers. In Germany, Georges Bose performed this feat with a circle of twenty strong, well-trained soldiers. All were helpless before the power of electricity.

Static electricity, however, dissipated rapidly. A show might come to a sudden halt, leaving the audience to wait impatiently while the scientist-showman recharged his glass tube or globe. Any device that could generate enough static electricity to last for a significant period of time would inevitably become huge and unwieldy. Many wondered if there was some way of storing a quantity of electricity in a moderately sized container where it could remain until used.

Pieter van Musschenbroek, a university professor in Leyden, Holland, thought he could solve this problem by charging a jar full of water. His first experiment with these jars gave him such a tremendous shock he almost gave up science entirely. Like other scientists, he thought that electricity was lost because it leaked through the glass walls of the jar like an invisible liquid. To keep it in the jar he coated the glass inside and out with a thin layer of tin (metal is an excellent conductor of electricity). Musschenbroek held his metal-coated jar in one hand while inserting a string charged with static electricity through the top of the jar with his other hand. All was well until the charge reached a crucial point and Musschenbroek found himself flat on his back with the wind knocked out of his lungs. Fortunately, he recovered and continued his research more cautiously.

In 1746, he sent a letter describing his jar to a friend in France. Soon scientists throughout Europe were using the so-called Leyden jars to store electricity. The jars were great fun. Jean Antoine Nollet, one of the chief lecturers at the French Academy, used Leyden jars to entertain the king. He reportedly made 180 members of the royal guard form a circle and leap into the air in unison by passing a charge through them via a Leyden jar. Louis XV was so delighted by this spectacle he ordered Nollet to repeat it with a record-breaking seven hundred monks. Nor were Leyden jars solely the property of the rich and famous. Glass and tinfoil had become relatively cheap in eighteenth-century Europe. The jars were simple to make and could be produced by almost anyone with the right materials and the ability to follow instructions. Over the next few years, Leyden jars appeared everywhere. People brought them out at parties to give guests a jolt. Every fair featured an electric show. Brave volunteers lined up to be shocked and then bragged about the nosebleeds, burns, and even concussions they endured. Electricity had once been deemed a dark and demonic force. Now it was just a trick.

Not everyone was pleased with this turn of events. Some of the more conservative members of society found these public performances ungodly and even blasphemous. They still believed there was something sacred in the natural world. Using electricity as a mere toy to entertain people might provoke divine punishment. Perhaps they even intimated that heaven would punish the sinners in the way deities were said to do from the beginning of time: God would strike them with a lightning bolt.

The scientists did not back down. If lightning was meant as punishment, they asked, why were churches among its most frequent targets? In America, Benjamin Franklin had noticed that churches with high pointed steeples were usually the tallest buildings in any village, city, or town. They were also the structures most like to be struck by lightning. Could these two facts be related? Though Franklin had little formal education, he was curious about everything. He kept up with the latest scientific developments abroad and had conducted his own experiments with Leyden jars. While others might regard electricity as a source of amusement, Franklin had a practical bent. He didn't know what electricity was made of, but he knew how it behaved. Certain materials could guide or conduct electricity along a path. If that path ended in the ground, the electric charge would fade away without doing harm. Even before anyone had determined that lightning was made of electricity, he had come up with a way to avoid it. In 1747, he described a device that would come to be known as the **lightning rod**:

> May not the knowledge of this power of points be of use to mankind, in preserving houses, churches, ships, etc., from the stroke of lightning, by directing us to fix, on the highest parts of those edifices, upright rods of iron made sharp as a needle ... Would not these pointed rods probably draw the electrical fire silently

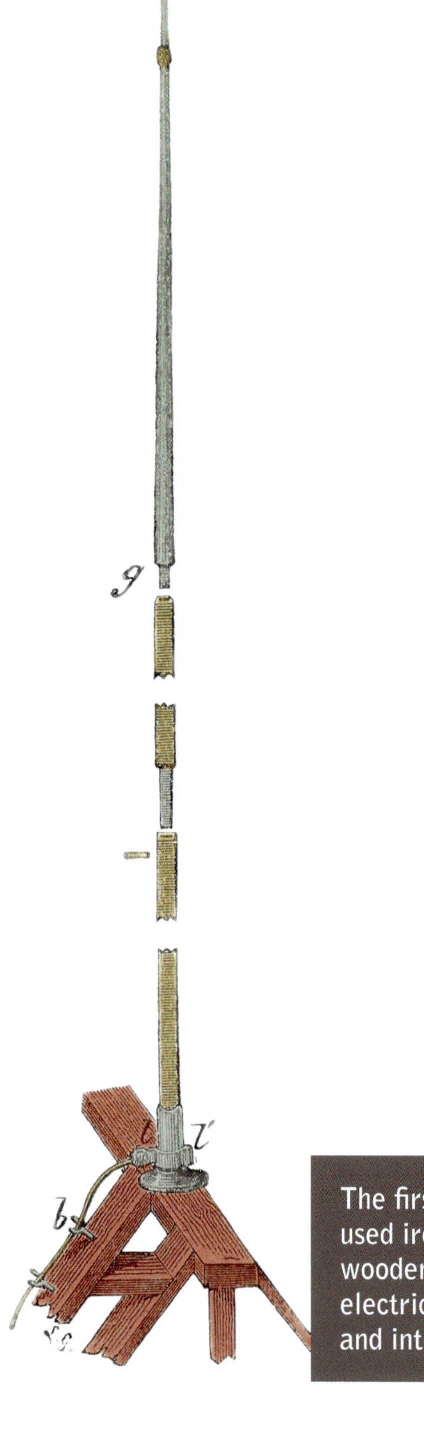

The first lightning rods used iron pikes inserted in wooden holders to conduct electricity around buildings and into the ground.

> out of a cloud before it came nigh enough to strike, and thereby secure us from that most sudden and terrible mischief!

As with many of Franklin's inventions, it's not clear when the lightning rod was first put to use. Franklin did install one in his own home and may have later designed one for Christ Church in Philadelphia after its new steeple was completed in 1754. The growing tensions between the British government and the American colonies caused Franklin to abandon science for politics. By 1760, he was no longer experimenting with electricity. A new generation in Europe, however, would continue to pursue the mystery of electricity. Within the next century, Luigi Galvani, Alessandro Volta, Andre Ampere, and George Simon Ohm would all give their names to the inventions and discoveries that revealed the nature of electricity to the world.

Lightning Strikes Again!

Despite the widespread use of lightning rods and other protective measures in modern buildings, lighting strikes still occur, though usually without serious consequences. The Empire State Building is struck an average of twenty-three times a year. Most people in the building do not even notice when it happens.

Some places are definitely more attractive to lightning than others. Lake Maracaibo in Venezuela holds the record. The largest lake in South America, Lake Maracaibo is located in a very thunderstorm-prone region. Lightning strikes the lake over six hundred times per square mile every year. It even has its own particular form of lightning, Catatumbo lightning, a type of continuous sheet lightning that can illuminate the sky for hours at a time.

In the United States, Florida, with its tropical climate, is lightning central, enduring slightly over twenty-five strikes per square mile every year. Louisiana is second with around twenty per square mile, followed by Mississippi with eighteen.

The most accurate data on lightning strikes is collected by the National Aeronautics and Space Administration (NASA) via the Tropical Rainfall Measuring Mission (TRMM) satellite. Launched in 1997 as a joint project by NASA and the Japan

Aerospace Agency, TRMM contains the Lightning Image Sensor (LIS), which records flashes of lightning as it circles the globe. In 2013 NASA published a world map showing the average number of flashes per year. Central Africa, Southeast Asia, and parts of South America experienced the most intense lightning activity. The satellite also contributes data to the Lightning Team at the Global Hydrology and Resource Center, which is dedicated to analyzing the connection between electrical storms, severe weather patterns, and climate change.

Italian physician Luigi Galvani's work on the nervous system of frogs introduced new concepts about the nature of electricity.

CHAPTER 2

Galvani and Volta: Revealing Electricity

Frogs are the unsung and uncelebrated workhorses of science. Easily bred and quick to grow to adulthood, they are excellent candidates for **dissection**, a fact that may have been first noted by the ancient Greek physician Galen in the second century CE. Forbidden from cutting open human corpses, Galen satisfied his curiosity by examining the innards of a variety of animals. Though his medical theories were not always correct, he did convince future physicians of the importance of learning from observation. Animal dissection became a tradition and a rite of passage passed down by students of medicine even during the darkest of the Middle Ages. It is not surprising, therefore, that the Italian physician Luigi Galvani had a ready supply of frogs in his laboratory in the late 1770s.

LUIGI GALVANI and the NERVOUS SYSTEM

Galvani had started his career as an obstetrician after receiving his medical degree from the University of Bologna in 1759. He soon discovered, however, that he was more interested in research than in treating patients and applied for a professorship at his alma mater. A few years later, in 1761, he became a lecturer in surgery and anatomy. Within a decade he had risen to the rank of full professor and was placed in charge of the university's medical laboratory.

Galvani was especially drawn to the study of anatomy. His medical education had given him some insight into the various organs and parts of the body, but, like every other physician of the era, he did not know exactly how the various systems worked together. The **nervous system**, in particular, was a mystery. Doctors had identified the larger nerves, and they knew that an electric current applied directly to a nerve of a freshly killed specimen could cause a muscle to contract. But they didn't know how the electricity might travel from the nerve to the muscle, and this is where Galvani's frogs come in.

Lucia Galvani Sparks Discovery

University science labs in the eighteenth century were often large, chaotic places with people conducting all kinds of different experiments at once. According to some accounts, Galvani was busy dissecting a frog one afternoon while a young colleague nearby worked with a machine that generated static electricity in glass globes. Much to Galvani's surprise one of the frog's legs jerked when he touched a nerve near the spine with a metal probe. The frog was not attached to the machine in any way. Curious, Galvani poked the frog again. Nothing happened, though the machine was still generating electricity. Galvani and his colleague no doubt spent a frustrating hour or two trying to figure out why the frog's body seemed to respond to the electric charge at some times and not others. Reportedly, it was Galvani's wife, Lucia, who first pointed out that the frog's leg was most likely to move if Galvani touched it with his metal probe at the very same moment that the machine gave off a spark.

Science labs filled with dead frogs were not exactly welcoming places for women in those days, but Lucia was a bit more educated than the average housewife. Her father, Gusmano Galeazzi, had been a professor of medicine and one of Galvani's early mentors at the university. She had grown up around scientific apparatuses and often assisted Galvani in the lab. She also edited his papers, an important job, for though Galvani was a dedicated researcher, he was a markedly poor writer.

Lucia's astute observation launched Galvani in an entirely new direction. Though he had little interest in electricity before, he now became obsessed with it. Once he had confirmed that electricity could pass through the air into the metal probe, from the probe to the frog's nerve, and from the nerve into the muscles, he began conducting a whole array of experiments to make dead frogs jump, twitch, and tremble alone and together. Eventually he realized he didn't even need the whole frog, just a leg with a nerve and muscle attached.

Frogs were slaughtered by the dozens. Galvani tied the flayed legs to an iron trellis outside his home and ran a copper wire upward from the fence to see if he could draw lightning down and make the legs jump on a stormy night.

Animal Electricity

So far the experiments had been interesting but not particularly revolutionary. Then Galvani made another discovery, the one that was to give him his place in history. The legs on the iron trellis did not need lightning to make them jump. On a completely sunny day he could make a frog leg contract by resting it against the iron structure and touching a brass probe to the exposed nerve at the same time.

He transferred the experiment indoors, placing a dissected frog on an iron slab and touching the nerves at various points with a brass knife. The combination of iron and brass seemed to generate some kind of electricity within the nerve itself. Lacking any other metaphor, Galvani likened the nerves to the Leyden jars so popular in electrical experiments. While the frog was alive, he hypothesized, electricity must be stored within the nerves much like it was stored in a charged Leyden jar, waiting to be discharged only when it came into contact with material that could serve as a conductor. The electricity remained in the nerves for a period of time after death and could likewise be discharged by the correct touch. He called this animal electricity.

Galvani was nothing if not a thorough researcher. For ten whole years, he conducted his experiments on animal electricity using various combinations of metals and other materials. He observed that though iron and brass were not the only two metals to produce the desired effect, he always needed two different metals and only metal would work. He wondered if the electricity he believed to be stored within nerves could somehow be used to bring dead bodies back to life. If this were true, then electricity was the very stuff of life itself. (Of course, he was wrong, and dead bodies cannot be reanimated by electricity.)

When Galvani, with his wife's editorial help, finally published a paper on his research in 1791, he made no claims regarding the potential of electricity to bring the dead back from the grave. In true scientific fashion, he laid out the details of his experiments concluding only that electricity was stored in the nerves and muscles and could be released by the application of two different metals.

The paper was written in formal academic Latin and did not attract a wide audience, though it did eventually add another word to the English language, **galvanize**, meaning to shock or jolt someone into action. More importantly, Galvani's theories of animal electricity were picked up again a century later and contributed to the development of the first **electrocardiograph** machine, a device that measures small electrical impulses released by the contraction of the heart muscles.

Galvani, however, enjoyed little fame during his lifetime. That does not mean all his frogs died in vain, though. His research paper caught the attention of Alessandro Giuseppe

Antonio Anastasio Volta, a popular and accomplished professor of physics at Pravia University in Lombardy, Italy.

BATTERY UP! VOLTA GENERATES ELECTRICITY

Unlike Galvani, Volta was not interested in biology or medicine. Frogs were safe around him for he had no use for dissection and regarded doctors as lesser scientists whose research was rarely as significant as that of physicists like himself. In fact, Volta may have ignored Galvani's paper entirely if his colleagues had not told him that Galvani had done some intriguing things with electricity. Volta was regarded as one of Europe's leading expert on electricity. He had started his career as a chemist and discovered the gas **methane** in 1776. His experiments with methane and other flammable gases led him to develop Volta's Law of Capacitance in the late 1770s. Capacitance refers to the capacity or ability of any body or object to store an electrical charge. Volta's law is based on his discovery that the electric capacity of a body and the electric charge it can give off are always proportional to one another. This in turn gave scientists a way to measure the force of electrical charges. In time, a unit of electrical force would become known as a volt.

Given that Volta was already well established in the field of physics, his opinion of Galvani's work mattered deeply to both Galvani and his fellow scientists. Volta dismissed Galvani's idea that electricity existed within nerves and muscles. To Volta, this notion smacked of ancient beliefs in magic, sorcery, and **spiritualism**. It was, in short, impossible. On the other hand, he was fascinated by how Galvani described using two separate metals to cause the frog's legs to jerk without the aid of any static electricity device. The electricity, Volta believed, was not within the frog's body but rather generated by some exchange or process that took place between the two metals. Ultimately,

The first electric batteries were made of metal discs separated by paper pads soaked in a chemical solution.

both Galvani and Volta were right in their own way. Small amounts of electricity are given off by the body, and electricity can be generated by the right combination of metals. Galvani's theories, however, languished for a long time, while Volta charged forward, inventing a working electric battery in 1800, less than a decade after he read Galvani's paper.

The first thing Volta did when commencing his research was literally stick out his tongue. Volta apparently had a very sensitive tongue. If two metals together could generate an electric charge, he reasoned, he should be able to feel a tingling sensation when he placed those metals on his tongue. To test this, he assembled a collection of small coin-sized disks of zinc, silver, copper, iron, and other metals and put pairs in varying combinations into his mouth. Zinc and silver seemed to produce the strongest sensation, though it was still very mild.

Interesting as this knowledge was, it didn't appear to be immediately useful. Volta's next question was how to harness the "electromotive force," as he called it, to produce a substantial and continuous electric current. Biology came to his rescue. This time it was a fish, not a frog, that prompted the needed scientific breakthrough.

Volta may not have had much interest in the study of animals but he did keep up with all kinds of scientific research. In 1797, the English naturalist and chemist William Nicholson had started publishing the *Journal of Natural Philosophy, Chemistry, and the Arts.* In one issue, he described in detail a dissection he had done of a torpedo fish known as the electric ray. The ray, Nicholson noted, had hundreds of special electric organs shaped like tiny discs on either side of its spine. These discs were packed into hexagonal columns that were separated from one another by thin sheets of cartilage. Though Volta did not believe in Galvani's theory of general animal electricity, he was well aware that special species of fish did generate electric charges.

The Battery Itself

Volta had a subscription to Nicholson's *Journal*, and Nicholson's article on the ray may have given him the inspiration he needed to build his first successful battery.

Volta made a stack of alternating zinc and silver discs, each separated by pieces of cardboard soaked in salt water. He ran copper wires from the top and bottom disc. When the disks were compressed a palpable electric charge ran through the wires. In this manner, Volta created a machine that could provide a continuous electric current.

Ever curious about the effects of electricity, Volta applied the electric wires to various parts of his body. Applying the current to his tongue produced a "sensation of light in the eyes," following by a prickling at the tip of his tongue. When he placed the wire in his ear he experienced "kind of cackling with shocks" so alarming he never repeated that particular experiment for fear that it might cause a permanent "shock in the brain."

Volta called the current generated by his battery "bimetallic electricity" because it was generated by the chemical reaction between two metals. He didn't know the nature of that reaction, but his realization that chemistry was involved in the creation of electricity was a great step toward the eventual discovery of the elusive electron many decades later.

Volta's Battery Becomes Famous

In 1800, Volta wrote a letter describing his battery to the head of the prestigious Royal Society in Britain who promptly published it in the society's journal. It's no exaggeration to say that the news of Volta's new device electrified scientific circles throughout Europe. At last scientists had a steady source of electric current to study, rather than just the sudden bursts given off by static electricity. For a while, some scientists thought that Volta had discovered an entirely new kind of electricity, something quite different from static electricity, and it took several years before that notion was dispelled by the later work of the mathematician James Clerk Maxwell.

The first scientists to construct batteries also hoped that they might find in them an endless source of electricity. They were inevitably disappointed when the batteries ran down, and electricity remained essentially mysterious. Nevertheless, Volta's battery gave them new ways to observe electricity at work, and it was a force to be reckoned with.

ØERSTED and RITTER: ELECTRICITY and MAGNETISM, TOGETHER AGAIN

Like many early nineteenth-century scientists, Hans Christian Øersted was more a generalist than a specialist. The son of a pharmacist in a small Danish town, he initially followed in his father's footsteps, receiving a pharmacy degree from the University of Copenhagen at the age of nineteen in 1796. Øersted, however, was more interested in the study of philosophy than managing a pharmacy shop. After working for one short year as a pharmacist in Copenhagen, he abandoned that career and returned to the university to pursue an advanced degree in philosophy, writing his thesis on the Danish philosopher Emmanuel Kant.

By the time Øersted had received a PhD in philosophy, Volta's battery had become well known throughout Europe. Øersted had built several, at one point demonstrating that electricity could be used to separate water into its two components: hydrogen and oxygen. Could electricity be the key to the great principle of nature he studied in graduate school, he wondered?

Øersted encountered others who thought it might, in particular the German physicist Johann Wilhelm Ritter. Like Øersted, Ritter had a number of wide-ranging interests. In

In the early nineteenth century, Danish chemist Hans Christian Øersted helped establish the connection between electricity and magnetism.

fact, Ritter was involved in the study of optics and light. He discovered the existence of ultraviolet light in 1801, the same year he met Øersted.

Ritter and Øersted spent a lot of time speculating on the nature of electricity. Ritter thought that magnetism and electricity might be somehow related. This was not a new idea, but a very old one. Ancient philosophers had assumed the two were related simply because an object charged with static electricity was capable of attracting other objects much like a magnet attracted metal. But there was little else to connect the two, and by the end of the eighteenth century the idea of any relationship had fallen by the wayside. It was well known that a lightning bolt or sudden discharge of electricity in the air might disturb the needle of a compass, pulling it from true north, but this was considered just another odd and capricious effect of electricity, not something worth studying. Scientists may not have been superstitious, but it did seem as if electricity often behaved like a kind of imp or poltergeist, causing sparks, jolts, and unpredictable sensations without much rhyme or reason. Scientists might observe electricity, but they still had little idea why it behaved the way that it did. Volta's battery, however, had begun to change that. With a steady supply of electricity, scientists could observe its effects over a given period of time. Both Ritter and Øersted wondered if the effect of electricity on a compass needle was not random but part of some larger system connecting electrical and magnetic forces.

Ritter was not inclined to investigate the matter in great depth, though. He was more interested in chemistry. He discovered a method of using the power from a battery to cover a wire with a thin coat of metal salts that had been dissolved in water, a process known today as **electroplating**, and he later experimented with building batteries using liquid chemicals rather than metal discs to conduct electricity, a process that eventually led to the development of **electrochemical** battery cells. Ritter was fascinated by the idea of opposing forces and reportedly made some investigations in the nature of Earth's magnetic field, but he died at the age of thirty-four in 1810 without publishing anything on this research.

Ørsted, meanwhile, had returned to Copenhagen, where he became a professor of physics in 1806. He was a popular teacher and liked to conduct experiments with his lectures. He remained convinced that electricity and magnetism might be related, and some of his experiments involved using an electric current running through a wire to move a magnetized needle. The needle, he noted, moved perpendicular to the wire when exposed to the current. This specific motion struck Ørsted as supremely significant. The needle did not jump or jerk randomly. Rather, it moved in an orderly and predictable way. If he ran the current through the wire in one direction the needle would turn left. If he ran it in the other, the needle would turn right. Clearly, there was a connection between electricity and magnetism and this connection was not random or sporadic. It was consistent, as if the needle was being pulled by an invisible magnetic field generated by the electric current.

After further experiments, Ørsted hypothesized that electricity flowing through a wire created a circular magnetic field around the wire, rather like an invisible spiral. When the current flowed in one direction this spiral rotated one way, and when the current flowed in the other, the spiral rotated in the opposite direction. In July 1820, Ørsted published a short paper on his observations entitled *Experiments on the Effect of a Current of Electricity on the Magnetic Needle*. Though the title was not very exciting the paper seized the imagination of Europe's scientific community. Public demonstrations of Ørsted's experiment with the needle soon followed. One such demonstration at the prestigious French Academy in Paris attracted the attention of a shy, introverted forty-five-year-old mathematician named Andre Marie Ampere.

AMPERE and OHM: MEASURING FIELDS of FORCE

Ampere had started life as something of a child prodigy. The son of a wealthy French merchant, he had taught himself basic math by the age of four and proceeded to read his way through his father's well-stocked library by the age of fourteen, teaching himself Latin along the way. He enjoyed the Enlightenment philosophers Rousseau and Diderot and had an avid interest in all fields of science, including astronomy, chemistry, and physics. But mathematics remained his great love.

The violent upheavals of the French Revolution brought his family's prosperity to an end. In 1793, the revolutionaries executed his father, to whom he was quite close. This tragedy triggered an emotional breakdown in the eighteen-year-old Ampere, and for nearly a year he ceased his studies and withdrew almost completely from the world.

Fortunately, he recovered his interest in mathematics and was able to earn a living by tutoring students in the nearby city of Lyon. Once the revolution was over, he moved to Paris and found a position as a lecturer at the Ecole Polytechnique, an institution that specialized in the growing fields of engineering, mechanics, and optics. Though he lacked a university degree, he was promoted to full professor in 1809.

By all accounts, Ampere was the classic absent-minded professor, shabbily dressed, dreamy, always with a piece of chalk in hand, and apt to write equations on any available surface. One anecdote tells of him suddenly writing a problem down on the back of a waiting carriage, only to go chasing after the carriage in a vain attempt to complete his problem as the vehicle rumbled away. His eccentricities, though, did not limit his career. Although he had never done any experiments with electricity, once he heard of Øersted's work, he devoted

Scientists have always sought ways to measure electricity. This device, called a Volt Ohm multimeter, indicates voltage, resistance, and amperage.

himself to the subject. He wanted to discover if there was any mathematical formula to explain the effect Øersted had described. In the course of his investigations, he noticed something Øersted had overlooked. Electric currents had a magnetic effect on one another. If two wires were placed parallel to one another and an electric current passed through each running in the same direction, the wires would repel one another just as two magnets repel one another from identical poles. However, if the currents ran in opposite directions, the wires would attract one another in the same way the negative pole of one magnet is attracted to the positive pole of another. In this manner, he established that electricity itself has magnetic properties. In 1825, he developed the mathematical formula that became known as Ampere's law. The law is complex and not easily summarized, but basically it states that the magnetic field generated by an electric current is proportional to that current. Its importance was immediately recognized by his peers, and Ampere was hailed as the Isaac Newton of electricity.

Almost equally important were Ampere's thoughts on what was going on within the wires when the electric current poured through them. Chemists had already established the existence of minute particles they called molecules. Now Ampere speculated that a special kind of charged particle he called electrodynamic molecules were responsible for both electricity and magnetism. An electric current, he stated, was generated by the movement of these particles. He could not prove the existence of these particles, though, and it would be another seventy-five years before the word "electron" was introduced to the scientific world. Today, an ampere (or **amp**) is a unit of measurement used to define the rate at which electrons flow through a conductor.

The same year Ampere developed his law, another mathematician, George Simon Ohm, was investigating electricity at the University of Cologne in Germany. Ohm was interested in the problem of resistance. Electric current did not always flow freely. Resistance might come in the form of a poor conductor or the close presence of insulating materials that inhibited the current and slowed it down. Like Ampere, Ohm thought that electricity was caused by the action between invisible particles. A volt, briefly described, is a unit of electrical charge. Ohm stated that an electrical current could be measured by voltage divided by resistance. One easy metaphor is to imagine someone riding a bicycle. The speed (current) of the bike is determined by both the strength of the peddling rider (voltage) and the amount of resistance the rider encounters, such as headwinds, traffic, or bumps in the road.

Ohm published his work on electricity, including the formula that later became known as Ohm's law, in 1827. Though his work did not attract much immediate attention, it gradually became highly regarded and essential to anyone studying electricity. In 1841, Ohm was made a member of Britain's Royal Society. By then Michael Faraday had devised the first electric generator and electricity had finally been put to good use in the form of the telegraph, an invention that would change communication forever. And an "ohm" would live on as the standard unit of electrical resistance recognized by scientists throughout the world.

Mary Shelley, Bringing Electricity to Life

One dark and stormy night in 1818, a young woman named Mary Shelley was sitting around a fire trading ghost stories with her husband, the poet Percy Shelley, and several of their friends. When it came her turn to entertain the company, she turned to science for her inspiration. The daughter of philosopher William Godwin and early feminist Mary Wollstonecraft, Shelley had enjoyed a wide and freewheeling education. She was familiar with Galvani's famous experiments on frogs and knew about attempts to reanimate dead bodies via electric current. With those experiments in mind, she wove a tale of an obsessive doctor who assembled a creature from butchered corpses and then brought his creation to life with a jolt of electricity in the form of lightning. Her listeners were so taken with the story they encouraged her to write it down. A few months later she published a short novel entitled *Frankenstein: Or the Modern Prometheus.* What experimenters had failed to do in reality, Shelley accomplished as fantasy. Electricity may not have engendered life, but it did help create a whole new genre of literature—science fiction. Shelley's novel has been the subject of dozens of films, television shows, radio dramas, stage plays, and even one ballet. The book has been in print continuously for two hundred years.

Born into a poor working-class family, Michael Faraday grew up to become one of Britain's most eminent scientists.

CHAPTER 3

Electromagnetism and the Age of Faraday

A twelve-year-old boy apprenticed to a London bookbinder hardly seemed destined for a great career in science. Born in 1797, Michael Faraday had grown up in poverty and received only a minimal education. He was a hard worker, though, and his master was apparently so pleased with his effort, he didn't mind when the young apprentice took time off to read some of the books he was binding. Faraday especially enjoyed reading about the latest scientific discoveries. When the binder got a contract to bind a complete edition of the new *Encyclopedia Britannica*, Faraday read through every entry regarding science and electricity. Another book he treasured was a small volume called *Conversations on Chemistry* written by Jane Marcet, the first English woman to write professionally on science. Marcet aimed her book at the general reader and wrote in a clear, conversational voice. The book included detailed descriptions of classic experiments in chemistry. Fascinated, Faraday decided to re-create them in order to test the results himself. He set about building his own equipment with whatever scraps of wood, glass, wire, and metal he could scavenge from the streets.

FARADAY'S PATH to SUCCESS

By his mid-teens, Faraday had assembled a working laboratory in his room above the shop and accumulated a small library of volumes on science. Marcet's *Conversations* always held a place of honor, though. He later credited Marcet with being the first to direct him toward a career in science, recalling that when he re-created the experiments in her books, "I felt that I had got hold of an anchor in chemical knowledge and clung fast to it."

As his interest in science increased, Faraday attended popular lectures given by the Royal Society where he took copious notes that he later bound into books for his own instruction. He also joined the City Philosophical Society, an organization of like-minded young men, along with a few women, who met to discuss science and natural history. Despite his progress, however, science was still more of a hobby for Faraday than a career. Lacking a university degree, he appeared to have little chance of landing a position in a real laboratory. When he sent a letter to the head of the Royal Society seeking employment, he received no reply. Science was an occupation for educated gentlemen, and the Royal Society had no impulse to hire a self-taught bookbinder no matter how well read and ambitious he happened to be.

Undeterred, Faraday sent out another letter, this one addressed to Sir Humphry Davy, the most famous chemist of the era. Faraday had enjoyed many of Davy's public lectures and included with his letter three hundred pages of notes he had made from the lectures and bound by hand. Faraday's packet arrived at the right moment, for Davy had just injured his eyes doing an experiment and was looking for someone to help him prepare a manuscript while he healed. He wrote to Faraday offering him a temporary job. A year later, in 1813, Davy helped Faraday secure a permanent position as a laboratory assistant at the Royal Society. In return for a modest salary, Faraday would sweep, clean, wash equipment, and do whatever general chores needed to be done. It wasn't a glamorous position, but it was exactly what Faraday wanted, for when he was not working he would be free to carry out his own experiments with the society's approval.

Faraday also benefitted from Davy's extended mentorship. He accompanied Davy on a tour of Europe, where he learned more about the work of Volta, Øersted, and Ampere. Upon returning to London in 1816, he began giving lecture-demonstrations on the subject to his old friends at the City Philosophical Society. Faraday had a passion for experimentation, and his happiest moments were in the lab surrounded by equipment. Experimentation, he believed, was the essence of science. In one of his early lectures at the society, he shared this idea with his listeners, telling them that a scientist should be,

> A man willing to listen to every suggestion, but willing to judge for himself. He should not be biased by appearances, have no favorite hypothesis; be of no school, and in doctrine have no master. He should not be a respecter of persons but of things. Truth should be his primary objective. If to these qualities be added industry, he may indeed hope to walk within the veil of the temple of nature.

Faraday's industry paid off. In 1821, he was named the superintendent of the house and laboratory of the Royal Institution, a position that came with a comfortable salary and the opportunity to devote himself almost entirely to his own goals. He was no longer Davy's assistant but his equal. Faraday would remain superintendent for the next forty-six years, during which he would completely transform the study of electricity and open the way for a new scientific age.

FARADAY'S DYNAMO: The FIRST ELECTRIC "MOTOR"

Once Faraday became familiar with the work of Øersted and Ampere, he wanted to explore the relationship between electricity and magnetism even further. If magnets and electricity shared so many properties, was it possible, he wondered, for magnets to generate electricity? The first answer to his question seemed to be a resounding no. Wrapping a copper wire around an iron magnet produced no electric current in the wire. Electricity had to be generated from an outside source like a battery. Nevertheless, Faraday persisted. He learned as much as he could about electricity and devised an almost endless array of experiments.

There are a number of stories regarding how Faraday first succeeded in generating electricity via magnetism. It's not absolutely clear when he did so, but around 1831 he made a very simple device by placing two bar magnets in a "V" connected at the bottom and running a copper wire between them at the top. He connected the wire to an apparatus he called a galvanometer, which used a needle to detect electric current. The needle remained still while the magnets sat in place, but whenever Faraday clapped the ends in and out, he noticed a very small but definite jump in the needle. It was not a steady motion but occurred only when the magnets were pulled apart and when they came back together. He also tried an even simpler experiment by placing a bar magnet within a copper coil, making sure the magnet did not touch the coil. As long as the bar remained at rest, the coil generated no electricity. However, when Faraday moved the bar back and forth rapidly within the coil, he detected a small electric current emanating from the copper coil. Faraday's use of magnets to generate electricity became known as electromagnetic **induction**. It was the biggest advance in the science of electricity since Volta's battery.

Faraday liked to give public lectures and demonstrations on electricity. His Christmas Day lectures for children were especially popular.

Faraday's next device consisted of a copper wheel placed on a stand between two magnets. When the wheel spun, it generated electricity from the magnets. A steadily spinning wheel could produce a steady electric current, making this the first electric **dynamo**, a forerunner of the electric motor. Faraday then went on to investigate the effects of electricity on different chemicals, creating the foundations for the modern science of electrochemistry.

While he conducted his experiments, Faraday remained a busy instructor and popular lecturer. As superintendent at the Royal Institution, he inaugurated a regular series of Friday lectures for the general public that still exist today, as well as an annual Christmas lecture for children. Guest speakers included some of the most eminent scientists of the age and covered topics from volcanoes to archaeology to the early theory of evolution. One of Faraday's own Friday presentations on electricity drew a crowd of over a thousand spectators.

Among the members of the audience was Ada Lovelace, the daughter of the famous poet Lord Byron and a gifted young mathematician. Lovelace was so enthralled by the lectures, she wrote to Faraday in 1844 asking to be his assistant, declaring she was ready to "bind my very life & soul to unwearied & undivided science." Faraday received her message kindly but turned her down, not because she was a woman but because his own failing health made it impossible for him to take on the responsibility of an apprentice. Lovelace later went on to work with Charles Babbage, the inventor of the first computer, and became known for her contributions as the world's first computer programmer.

Faraday continued to work at the Royal Institution almost until the age of sixty-seven. He developed into a beloved, though sometimes controversial, public figure. His annual Christmas lectures for children inspired him to become an advocate for science education. He was one of the first to press for the inclusion of science courses in the school curriculum and insisted that all students, not just those at wealthy private institutions, should learn scientific methods. He was also dismayed at the pollution of the River Thames in London and became an early crusader for water purification. At his death, he was honored as Britain's greatest scientist since Sir Isaac Newton.

JAMES CLERK MAXWELL and HIS GREAT EQUATIONS

Though Faraday was a brilliant scientist, he was not a gifted mathematician. Faraday knew his discovery of the fact that electricity and magnetism acted upon one another with what he called "fields of force" could be expressed in mathematical terms. He just didn't know how to do it. "If I could live my life over," he once wrote to a fellow scientist, "I would study mathematics. It is a great mistake not to do so, but it is too late now." Fortunately, Faraday had help. Much as Ampere had translated Øersted's work into equations, a young Scottish mathematician, James Clerk Maxwell, stepped forward to make mathematical sense of Faraday's electromagnetic dynamo.

The two men were born almost a generation apart; Maxwell came into the world in 1831, the very year Faraday had made his great discovery that magnets could generate electricity. Maxwell's family exposed him to the study of science early in life. His father, an Edinburgh lawyer, often took him to scientific lectures at the University of Edinburgh. A letter written by his mother reveals that even at the age of three he pelted his parents with the question, "Show me how it doos?" as he investigated locks, doors, keys, and what his mother called "bell wires," which the family pulled to ring for servants. "He drags Papa all over to show him where the holes go through," she added. As he grew older, Maxwell roamed the fields gathering specimens of rocks, plants, and bones for his natural history collection, and he covered his schoolbooks with all kinds of diagrams for imaginary machines. Geometry

Scottish physicist and mathematician James Clerk Maxwell developed four mathematical equations to explain the nature of electromagnetic energy.

was among his favorite subjects, and at the age of fourteen, he published a short paper on his own method for drawing a perfect oval with the use of a string.

At sixteen, he graduated from the Edinburgh Academy with high honors in both mathematics and English literature. For three years, he took classes at the University of Edinburgh, wavering between a career in law and science. There were few full-time jobs for scientists in nineteenth-century Britain, and obtaining a post was highly competitive. Law promised stability and security. Maxwell's own father loved science but had chosen the better paying legal profession. Science, however, promised a life of intellectual adventure Maxwell believed he could not find elsewhere. He had little patience for the formalities of dress and manner that would be expected of a lawyer. His friends noted that he dressed plainly without "the vanities of starch and gloves" and preferred to travel like a common workman in third-class train carriages, claiming that the hard seat helped keep his mind alert.

A Scientific Life

Once Maxwell had decided to pursue science he acted with characteristic self-discipline. Upon enrolling in Cambridge University in 1850, he set a rigorous schedule for himself, rising at 5:45 for an hour or two of outdoor exercise, either swimming, rowing, or hiking, then reading a selection from a Greek or Latin classic before settling down to his studies for the day.

His efforts gained him the success he craved, for he won several mathematical competitions, and, in 1854, became a Fellow of Cambridge University, a highly coveted appointment among young scholars. His interests ranged far and wide, and he developed mathematical formulas for all kinds of everyday phenomena from how a piece of paper twirled in the wind to how cats righted themselves as they fell through the air.

"The Rule of Plan," he wrote to a friend, "is to let nothing be left willfully unexamined."

Given his curiosity, it seems almost inevitable that Maxwell would become intrigued by electricity, the greatest mathematical puzzle of his age. Another mathematician, William Thompson, had tried to explain Faraday's "fields of force" without quite succeeding. Maxwell read Thompson's work and even corresponded with the older scholar, but rather than build on it, he decided to go back to the beginning and read all Faraday's original papers himself. In Faraday, Maxwell found something like his other half. While Maxwell was almost solely interested in working out mathematical formulas in his head, Faraday was completely devoted to concrete, physical experiments.

"It was perhaps for the advantage of science," Maxwell later wrote, "that Faraday … was not a professional mathematician." Faraday's very lack of mathematical skill freed him to conduct so many different experiments and record his results in what Maxwell praised as "natural, untechnical language."

Maxwell's Early Ideas About Electricity

One of the first ideas that Maxwell dismissed after reading Faraday's papers was Thompson's belief that electricity and magnets acted upon one another "from a distance." It was commonly held theory that because an electric wire and a magnet did not touch one another, something had to leap across the space between them in a single motion, much like a flea might leap from a dog to a man and back again. This was not surprising, given that most scientists regarded space as simply empty, motionless air. But Maxwell had studied optics and had already encountered the idea that light moved through space in waves. Perhaps, he hypothesized, space was not empty but filled with all kinds of waves carrying light, sound, the forces of gravity, and ultimately electricity and magnetism, too.

In 1857, the twenty-six-year-old Maxwell published a paper entitled "On Faraday's Lines of Force." In it he presented the idea that electricity and magnetism traveled in waves, similar to an invisible rope snaking either horizontally or vertically, that interacted with one another as they traveled. What's more, these waves engendered more waves. Magnets could generate electricity and electricity could generate magnetism. Each fed off the other, carrying the current onward. During the 1860s, Maxwell expressed his insights into four mathematical equations that came to be known as Maxwell's equations.

The Equations

Maxwell's equations presented four fundamental principles of electricity and magnetism and proved that static electricity, electricity generated by batteries, and that of electromagnetic dynamos were the same thing. All these forms of electricity were governed by the same mathematical principles. There are many ways of summarizing these principles, some far more complex than others. One of the simplest is that Maxwell's equations enabled scientists to calculate the size of the electric field created by an electric charge, the strength of electric and magnetic waves, and how fast those waves could travel. These basic equations helped later researchers in their efforts to develop more powerful electric generators. They also helped predict the existence of radio waves and laid the basis for Albert Einstein's theory of relativity. In 1873, Maxwell compiled all his work on electromagnetism in a single volume, *A Treatise on Electricity and Magnetism*, which is still considered one of the classic works of science today. Writing on Maxwell in 1964, the preeminent physicist Richard Freyman said simply, "He brought together all the laws of electricity and magnetism and made one complete and beautiful theory."

JOSEPH HENRY RINGS a BELL: The STUDY of ELECTROMAGNETISM in AMERICA

Europe may have been the center of scientific inquiry in the nineteenth century, but America was not the intellectual backwater many Europeans thought. At the very time Faraday was building his dynamo in London, a schoolteacher in Albany, New York, was building his own electromagnetic "motors."

Born in 1797, Joseph Henry had already enjoyed an eclectic career as a silversmith, watchmaker, storekeeper, surveyor, and actor before becoming a mathematics instructor at the Albany Academy in 1826. He had been quite serious about pursuing a life on the stage, until a book entitled *Popular Lectures on Experimental Philosophy* happened to draw him toward science. He enrolled in the academy as a student and after a few years was asked to become a teacher.

Henry had a practical bent and liked the idea of **mechanical engineering**. The Albany Academy did not have a large library, but the school did subscribe to some of the more important British scientific journals. When he encountered articles about Volta's battery or Faraday's dynamo, his immediate reaction was something along the lines of, "Well that's very interesting, but what good does it do?"

Electricity may have had great scientific value, but it still had very little practical use. Scottish inventor James Watt had patented the steam engine in 1781, and the Western world was rapidly entering the great age of steam. In 1807, Robert Fulton had introduced the world's first commercial steamboat in Albany. A few years later, in 1811, John Blankinsop of Britain demonstrated the first steam railway. Steam powered not just transportation but factories of all kinds. The Industrial Revolution was driven by steam. Electricity was hardly more than a scientific curiosity far removed from the work of everyday life.

American engineer Joseph Henry experimented with electromagnetic devices for lifting large objects using iron magnets wrapped in copper coils.

Shortly after he started teaching, however, Henry read an article about William Sturgeon, a scientist in Britain who had wrapped a copper wire around a large chunk of iron and attached the wire to a voltaic pile, or battery. The electricity magnetized the iron, which then picked up any metal objects within its reach. This experiment was not terribly unusual. In fact, it was a regular part of lecture demonstrations throughout Europe. What fascinated Henry, though, was the question of strength. How much weight could an electromagnet lift, he wondered.

With the help of his teenaged students he built a battery, attached it to a wire-wrapped piece of iron, and watched as the new magnet hoisted 9 pounds (4.1 kg). Soon it became clear to him that more wire meant more strength. He built electromagnets that could lift 20, 30, 100 pounds (9.1, 13.6, 45.35 kg). Finally, he encased a small block of iron in so much copper wire that when he switched the battery on it lifted several blacksmith anvils totaling around 1,500 pounds (680.4 kg). The device was certainly entertaining, for when Henry shut off the battery the anvils came crashing down much to the delight or terror of the onlookers.

Henry Makes His Mark

Henry might have been confined to demonstrating his wonder at county fairs, but fortunately he had attracted some serious scientific attention, and by 1830 he was making electromagnets for the professors at Yale. Sometime between 1830 and 1832, Henry also discovered the same principle of electromagnetic induction as Faraday. Because Faraday published his findings first, he was hailed as the father of electromagnetism. Henry did not dispute this and refused to become bitter over being considered the runner-up. The two men met during a short trip Henry made to London, and he described their relationship as cordial. Besides, Henry's own endeavors were enough to give him lasting fame. His work inspired John Davenport of

Vermont to create the first electric motor in 1834. He also contributed to the understanding of electromagnetic induction that later helped develop **transformers** and generators.

Revolutionizing Communication

One of Henry's greatest discoveries occurred when he started to think small, not big. He had started to investigate how far an electric current could flow through a wire. He ran a wire between two buildings. At one end, he had a battery. At the other, he placed a piece of iron with a small scrap of metal beside it. When an assistant at this end saw the scrap pulled to the magnetized iron, he shouted through the window to Henry that the electric current had succeeded in moving the length of the wire. At some point, the assistant recommended placing a metal bell beside the iron. The magnetized iron pulled the bell's clapper, causing it to ring, and the assistant didn't have to shout any more. Henry unrolled greater and greater lengths of wire until he was able to ring a bell electronically from a distance of slightly over a mile. His reports on this research attracted the attention of an artist turned inventor named Samuel Morse. Electricity, Morse realized, might not be able to power factories, ships, and trains like steam, but it could do something that steam power could not—it could communicate. Electricity was about to be put to its first practical use. People called the new invention the telegraph.

Smithson's Gift to America

Though the wealthy English chemist James Smithson never set foot in the United States, he made a lasting impression there. When he died in 1829, he left his money to his nephew, stipulating that if his nephew had no children the nephew would pass the money on to the United States in his own will. In 1836, Smithson's childless nephew died, leaving approximately $500,000 to the US government for the purpose of creating "at Washington, under the name of the Smithsonian Institution, an establishment for the increase and diffusion of knowledge."

It took ten years for Congress to draw up the necessary paperwork. On August 10, 1846, President James Polk signed an act of Congress formally establishing the institution. The act called for the institution to be governed by a board of regents. One of the first things the board did was ask Joseph Henry to become the Smithsonian's first superintendent or secretary. As overseer of the new institution, Henry kept track of the growing collection of specimens and scientific instruments that poured in from donors. He organized a nationwide network of weather observers that later became the National Weather Service and offered advice and encouragement to Alexander

Graham Bell, who came to discuss his idea for an electrical voice apparatus, or "telephone," in the spring of 1875.

Henry died in 1878. In 1882, a statue of him was dedicated on the National Mall in front of the Smithsonian Castle. The American composer John Phillips Sousa created a special march for the occasion.

American inventor Thomas Edison developed the first long-lasting electric lightbulb at his laboratory in Menlo Park, New Jersey.

CHAPTER 4

Technology and Science: Electrical Inventions and Discovery of the Electron

One of the most common experiments in any elementary school science class is the potato battery. Take a potato (or any other slightly acidic fruit or vegetable), and insert a zinc nail in one end and a copper nail in the other. Run a wire conductor between them, and the potato will generate a tiny electric charge. This occurs because the acid in the potato dissolves minute particles of zinc that release electrons. This chemical reaction gives the zinc a negative charge. At the other end of the potato, the acid has the opposite effect on the copper nail, enabling it to absorb the loose electrons released by the zinc and acquire a positive charge. In technical language, the zinc and copper are the **electrodes**. The zinc is the negative electrode, or **anode**; the copper is the positive electrode, or **cathode**. The process of the electrons going from negative to positive poles generates an electric current.

A battery made from a potato or other acidic vegetable can generate enough electricity to power a small digital clock.

The electron was the missing piece in the great puzzle of electricity. The discovery of its existence illuminated the mystery of static charges, explained how Volta's battery worked, and enabled scientists to comprehend the nature of electromagnetism. Yet though the electron was the key to all these things, it wasn't needed to open the doors to electricity's use. Technology far outstripped science when it came to putting the electron to work. Telegraphs, lightbulbs, phonographs, and telephones were all developed by people who had little knowledge of chemistry or physics. The existence of the electron itself was not verified until the last years of the nineteenth century, and by then the second great industrial revolution, the great age of electricity, was well underway.

"PARTICLES EXTREMELY SUBTILE": FRANKLIN and STATIC ELECTRICITY

The idea that electric charges engaged positive and negative poles goes back to Benjamin Franklin's early work with static electricity. Franklin theorized that when someone charged a glass tube the tube acquired an excess of something he called electrical "fluid." If one individual charged himself by holding the tube and then passed it to another person, the tube would be taking some of the electrical fluid from the first person and transferring it to the other, who would then have an excess of this fluid. When the two people touched their knuckles together, they would generate a spark, releasing this fluid. Franklin believed this spark meant that the fluid was returning to its original balance passing from the second person back to the first. Franklin called the first person the "positive" pole and the second the "negative" pole. He was correct in assuming the transfer of electricity involved two poles, but later researchers reversed the negative and positive poles. More importantly, however, Franklin realized that the glass tube did not generate electricity itself; the tube merely conducted electricity from one

Benjamin Franklin was one of the first people to postulate that electricity might consist of minute invisible particles.

pole to the other. Electricity, he stated, existed in all things. We don't perceive it when it is in balance, or neutral. Only when an imbalance occurs, do we notice electricity.

Volta had not yet invented his battery, so Franklin was limited to experimenting with static electricity and Leyden jars. People during his era spoke of "storing" electricity in Leyden jars. But was the electricity really inside the jar itself, he wondered. He determined it wasn't by inserting and removing various objects from charged jars, making sure they did not touch the bottom or the sides of the jar in any way. If the inside of the jar was filled with electricity, he should have received a shock from these objects, but he didn't. The air inside the jar was not filled with electricity. Next he worked with water. Water was a well-known conductor of electricity. Yet when he charged a jar filled with water and poured that water into a second jar, a wire inserted into the water-filled second jar delivered no shock. The electricity, Franklin concluded, was not in the water. Otherwise, he would have been able to transfer electricity via water from one jar to the other. The glass surfaces of the jar alone contained electricity.

Soon afterward he began thinking of electricity as consisting not of fluid but of particles. These particles were so tiny they could move through solid material. Here, in Franklin's work, was the first inkling of an electron. In 1750, he wrote to his friends in London describing his experiments. Franklin used the metaphor of a sponge. The sponge filled with particles of water, he explained, is like a neutral body that has neither a positive or negative charge. If the sponge absorbs too much water, it must give some off in order to return to its neutral state. If the sponge loses water and begins to dry out, it will naturally try to gather more back in. The English scientist William Watson introduced Franklin's findings in the journal of the Royal Institution, telling readers that,

British physicist John Dalton developed the first comprehensive atomic theory. His work led to the later discovery of the electron.

> This ingenious author, from a variety of curious and well-adapted experiments, is of the opinion that the electrical matter consists of particles, extremely subtile; since it can permeate solid matter, even the densest metals, with such ease and freedom as not to receive any perceptible resistance.

Franklin's subtle particles would continue to intrigue investigators though it would take over a century for them to finally become known as electrons.

"CURIOUSER and CURIOUSER": The INVISIBLE WORLD

In 1865, Lewis Carroll published a story about a little girl who grew or shrank at alarming rates as she sampled various cakes, goodies, and other "stimuli" in a wondrous underground world. "Curiouser and curiouser," Alice would exclaim as she changed size once more. *Alice in Wonderland*, of course, is fantasy not fact. But Lewis Carroll was the pen name of Charles Dodgson, a clergyman and mathematician educated at Oxford University, and Dodgson was no doubt aware that scientists at that time were discovering the universe around them was both much larger and much smaller than anyone had ever imagined. Powerful new telescopes had revealed distant stars and meteors, as well as the planet Neptune, detected in 1846 by astronomers John Couch Adams and Urbain Le Verrier. At the same time, physicists were learning that the universe around them was made of particles much smaller than anyone had ever thought possible.

The word "atom" is derived from the Greek word for "indivisible" or "uncut." Around 400 BCE, the Greek philosopher Democritus hypothesized that the universe consisted of two things: atoms and empty space, or "the void." Atoms, he believed, were the infinitesimal but solid bits of material out of which everything was made and they could never be destroyed. Historians often quote him as saying, "Nothing exists except atoms and the void. All else is opinion." Atoms, however, were regarded as a merely abstract or **metaphysical** idea for several centuries until the French astronomer and philosopher Pierre Gassendi took a closer look at the classical idea of **atomism** in the 1640s.

The development of the microscope in the Netherlands in the 1620s had allowed people to observe things too small to be seen by the naked eye for the first time. Everyday materials were revealed to consist of tiny cells. People could not see atoms, but if cells could exist, it seemed at least plausible that even smaller particles might be found within cells, and within those particles something even smaller yet. With this in mind, Gassendi began to think in terms of molecules and atoms. Molecules, he hypothesized, were formed by groups of atoms, and the different types of molecules gave each type of matter its particular qualities. Gassendi did not know how atoms combined to make molecules, but his theories were useful to the English chemist John Dalton, who began to investigate the composition of gases in the early nineteenth century.

Dalton developed what historians now recognize as the first systematic atomic theory. Like his predecessors he believed that atoms could not be divided. This, of course is false, but Dalton also presented other hypotheses that proved to be true. Among them was the idea that each **element** consisted of identical atoms. For instance, only oxygen atoms make up oxygen. Atoms that join together to produce molecules do so in a constant proportion to one another. Water, for example, always consists of one part oxygen to two parts hydrogen. In addition, Dalton stated, these proportions can always be expressed in small whole numbers. This last point became known as Dalton's law when he published it in 1808 in his groundbreaking volume, *A New System of Chemical Philosophy*.

American telegraph inventor Samuel Morse developed Morse code, the first universally used code for electronic communication.

Although Dalton's theories were not immediately embraced by those scientists studying electricity, the idea that these tiny particles called atoms might be important did begin to take hold. Dalton also investigated atomic weights and was the first to create a system for recording those weights. He was an honored speaker at Faraday's Friday lecture series at the Royal Institution, and despite the difficulty of his subject matter, he became something of a scientific celebrity. When he died at the age of seventy-seven in 1844, nearly forty thousand people came to view his body as it lay in state.

CODES and CABLES: The INVENTION of the TELEGRAPH

On May 24, 1844, just a few months before Dalton died, a former artist named Samuel Morse sat down at a telegraph machine in a chamber of the Supreme Court in Washington, DC, and sent the message "What hath God wrought?" to a railroad operator 40 miles (64.37 km) away in Annapolis, Maryland. Morse knew nothing about atoms and wouldn't have considered them important if he did. Like many of his fellow inventors, he was solely interested in using electricity, not analyzing it.

Starting with the invention of the telegraph, electrical technology began to diverge from electrical science. By the early twentieth century, an electrician and a physicist would occupy two separate worlds. The electrician was a tradesman, a skilled laborer, who kept electrical machines operating. A physicist, for the most part, held an advanced degree and worked at a university. Yet technology and science can be intertwined, and the electrical inventions of the last half of the nineteenth century eventually spurred greater insight into the nature of electricity itself.

The idea of using electricity to communicate had been around for at least a decade before Morse sent his famous message. During the 1830s, a number of scientists in Germany had experimented with various ways of sending messages through wires. The great problem was devising a system of signals that could be used easily and efficiently. Most of the devices were complicated and cumbersome. One used twenty-six separate wires plus nine more for each numerical digit. The sender had to use the correct wire for each letter or number. Other systems used one or more meters with needles that pointed to a certain number or letter depending on the strength of the electric charge. It was hard to control the charge, and the needles were not always accurate, causing most messages to be fatally garbled.

Cooke, Wheatstone, and the Needle Method

Despite its flaws, the needle method was used with some success in England by the team of William Cooke and Charles Wheatstone. Cooke was a businessman-inventor and Wheatstone a physicist known for his experiments with sound and light.

Wheatstone's father and uncle had both been professional musicians, and although he did not follow that career, music inspired his first works. In 1821, he amazed the public with his "Enchanted Lyre," an instrument played by an "invisible" musician. In reality, the music was created by vibrations sent through a hidden system of steel rods in the walls controlled by a person in another room. He went on to invent several other instruments, among them the Wheatstone Concertina, a combination of harmonica and accordion. A portable harmonium, or pipe organ, won him a prize at England's Great Exhibition of 1851. In 1834, he turned his attention to the study of light and conducted a series of experiments to measure the speed of electricity using a system of rotating mirrors. This achievement won him an appointment in the Experimental Physics Department at prestigious King's College in London.

Wheatstone was aware of experiments with telegraphy, but he had expressed no interest in it himself until he received a visit from William Cooke in February 1836.

Cooke had dropped out of medical school to wander around Europe, taking in the sights and attending public lectures. Sometime in the early 1830s, he saw a demonstration of a telegraph device in Germany. Around the same time, engineers in Britain had started to develop the first railway lines. Cooke ingeniously put the two together in his mind. Railroads could use a quick method of relaying messages, and what's more they might pay for it. Cooke initially presented his idea for a railway telegraph to the great Faraday, but Faraday seemed reluctant to become involved in a business venture. He did, however, refer Cooke to Wheatstone, whom he knew and admired.

Like Faraday, Wheatstone was initially against mixing commerce with science. Cooke's promise of fortune finally swayed him, though, and the two men set to work. Wheatstone perfected the device while Cooke raised funds. In 1838, they won a contact to install a telegraph for the Great Western Railroad, which ran a line from Paddington Station in London to towns north and west of the city. It was the world's first commercial telegraph. The machine was powered by batteries. It employed several wires and used five different needles to point to numbers and letters on a large board. Later, Wheatstone invented a rotating needle on a single round head that did away with the complicated board and enabled operators to send messages with a single wire.

Wheatstone and Cooke's telegraph received a tremendous boost in the press when it proved instrumental in apprehending criminals. In one case, a telegraph operator at Paddington Station learned that a well-known pickpocket had just been seen boarding a train. The operator notified the next station, and the police were ready and waiting to seize the unlucky thief who had just lifted a lady's purse when the train pulled in. On a more somber note, in January of 1845, a woman was murdered in the town of Slough on the train line. The police believed the suspected murderer had fled to London. They telegraphed a description of the man to Paddington Station and he was arrested a few days later.

Despite these triumphs, the telegraph did not thrive. Wheatstone and Cooke quarreled over money, and when they could not come to an agreement they split up and their company disbanded. By then, their system of meters and needles had been superseded by a new streamlined method of sending messages using nothing but dots and dashes. It was called Morse code.

The CODE THAT CHANGED EVERYTHING

According to almost every biographer, Samuel Morse was a difficult and cantankerous man. He expended a great deal of time and energy on seemingly endless lawsuits and patent fights, often taking credit for inventions that were not his own and cheating his business partners out of their rightful recognition and money. He defended the practice of slavery and despised the Irish immigrants pouring into the United States in the early nineteenth century so much he joined the anti-immigrant Nativist Party and ran as their candidate for mayor of New York in 1836. (He lost by an overwhelming majority.) In his twenties, he had aspired to become an artist and portrait painter, but although he had talent, his work did not sell. His failure in the art world rankled him for the rest of his life.

He did have one good idea, though. Returning from a three-year sojourn in Europe in 1832, he made friends with a fellow ship passenger who knew something about electricity and the attempts Europeans had been making to create a telegraph. Morse was forty-one and a professional, artistic, and financial failure. This telegraph idea seemed to have possibilities. He put away his paints and easel and tried to think of how an electric wire might send a code. His friend had told him that an electromagnet at the end of a wire made a kind of tap or click when it was activated. Morse's first idea was to assign a discrete set of short and long clicks to every word in the dictionary. This soon proved absurd. Then it dawned on him that the clicks did not have to relate to words, only to letters. Morse was never clear on exactly how he devised his code, but he later claimed that by the time the boat docked in New York he had come up with a system that assigned a combination of short and longs clicks, which he called dots or dashes, to each letter of the alphabet, plus an additional combination for every digit from zero to nine. This was a total of only thirty-six unique combinations of dots and dashes. With this code, he was convinced anyone could send a message of any length.

He spent the next ten years learning as much as he could about electricity, signing on business partners, and trying to convince the US government to fund his work. In March 1843, Congress named him superintendent of the United States telegraph system and awarded him $30,000 to construct a telegraph line from Washington, DC, to Baltimore. A little over a year later, he gave his first public telegraph demonstration.

By 1861, transcontinental telegraph lines snaked across the country from New York to San Francisco and every place in between. In 1870, Cyrus McCormick completed laying the first transatlantic telegraph cable joining North America with England. By 1880, there were nine such cables between the Americas and Europe. Telegraph wires connected the people and nations of Europe, too, and no matter what language people spoke, they all used Morse code. In the United States, the Western Union telegraph boy on his bicycle became a familiar sight as he delivered telegram messages in nearly every city and town.

The telegraph operated twenty-four hours a day. For the first time, newspapers could publish stories almost as soon as events in faraway places unfolded. Corporations grew as communication with remote offices became easier. There were even rumors about young people flirting with one another via telegram, a shocking development in the prim Victorian Age. Yet big as these changes were, electricity had just begun to transform the world. Electricity had proven it could communicate. Now, under the guidance of Thomas Alva Edison, it would become light.

EDISON and TESLA: DC and AC

Perhaps no name in history is as closely associated with electricity as that of Thomas Edison. In his lifetime, he would take out a total of 1,093 US patents, the vast majority of them for devices related to electricity, either new inventions or improvements on existing ones. Edison was more than a prolific inventor; he was the consummate entrepreneur. He wanted to make money, and he wanted to make a lot of it.

Born in 1847, he grew into shy, awkward child who never did well in school. He loved to read about science, though, and spent most of his time at home concocting chemistry experiments in the basement. At the age of twelve, he built his own telegraph set. Four years later, at sixteen, he left home to become a professional telegraph operator, a job that allowed him to travel from city to city across the United States, tinkering with inventions along the way. At twenty-two, he moved to New York and tried to develop an electronic vote-counting machine for the local legislature. Politicians in the city's corrupt Tammany Hall swiftly vetoed that idea. Edison found more success on Wall Street, where he developed a device to help stock market tickers function more smoothly and a telegraph that could print messages as they came in. This netted him and two business associates $15,000. After that Edison was on his way.

In 1875, be built an enormous shed in Menlo Park, New Jersey, that he called his "invention factory." There, he and his crew of workmen churned out hundreds of inventions, large and small, ranging from an artificial perfumed rose to the first phonograph machine. The biggest challenge, however, was light.

The Lightbulb

Researchers knew that if someone inserted an electric wire attached to a filament into a sealed glass tube or globe, the electric current would cause the filament to **carbonize** and glow, creating a radiant light for a few minutes until the filament was destroyed. Edison was convinced he could create a light that lasted several hours or more. He was not the only one trying to solve this problem. In fact, there was an international competition on to see who could develop the world's first practical lightbulb. Edison threw all his money, attention, and ingenuity into the race.

First, he pumped more air out of the glass bulb before sealing it to help slow down the **oxidation** of the fiber. Next he began testing fibers to see which would last the longest. He tried every material he could get hold of, from cardboard soaked in sugar to human hair, all with varying amounts of success. By 1880, he had created bulbs that could last anywhere from 40 to 550 hours. In January of that year he received US Patent No. 223,898 for an "electric lamp" that consisted of "a carbon filament or strip coiled and connected … to platina contact wires."

Edison and his crew continued to perfect the bulb. A few months later they were up to test number 1253, using a filament of bamboo pulled from a nearby fan. The bulb burned and burned and burned for a record 1,200 hours. Edison was in the lightbulb business for good. In 1882, his factory manufactured over a thousand lightbulbs a day, and he had received a commission to create an electric grid in New York City that would bring his lights to thousands of homes and offices.

Then he ran into a major roadblock. An electric current running in one direction weakens as it passes through a wire.

This type of current is called **DC**, or direct current. The longer the wire, the weaker the current will be at the terminal end. The inventors of the telegraph had solved this problem by placing small battery sets called amplifiers along their wires to boost the signal. As the current passed the amplifier, it would cause the amplifier to switch on and give the telegraph signal the necessary increase in power it needed to reach its destination. The telegraph system, however, ran only from one operating station to another. Edison wanted to wire an entire city. In addition, he didn't rely on batteries but on electromagnetic generators powered by steam. If he wanted his current to work, he would have to install hundreds of these generators throughout the city, placing a huge, chugging, steam-spewing machine at just about every crossing and corner.

TESLA INVENTS and INTERVENES

The solution to Edison's problem came from the Serbian mathematician and engineer Nikola Tesla. Born in 1856, Tesla was a strange, dreamy, intense man who claimed he could visualize his inventions entirely in his mind before he built them. While Edison was a hands-on tinkerer, Tesla worked behind closed eyes and then constructed a machine exactly as he envisioned it, down to the smallest details.

At the age of twenty-five, Tesla had invented an electromagnetic generator that used **AC**, or alternating current. Alternating current means that the current switches directions, flowing from positive to negative poles and then reversing course to flow in the opposite direction. Tesla's machine used three magnetic copper coils. As the electric energy in one coil reached its peak pulling the current in one direction, it would trigger a reaction in the next coil, which then sent the current in the opposite direction.

Tesla had heard about Edison's enterprise in New York, and in 1884 he arrived in the city to see if the famous inventor was interested in his idea. Edison wasn't. He thought alternating current was useless and possibly dangerous. Nor was he alone in this assumption. Most people still pictured electric current as some kind of river: in order for a boat to reach its destination, the river must flow smoothly in one direction. Alternating current was like a series of powerful, stormy waves on the river, tossing the boat back and forth and preventing it from moving forward. Tesla, however, knew this metaphor was false. He was convinced alternating current was cheaper, safer, and more effective than direct current. He just wanted a chance to prove it.

Edison refused to even try Tesla's device. The two met briefly only once and did not like each other very much. Edison was notoriously indifferent to his physical appearance, often going for days without changing his clothes or combing his hair. Tesla was immaculately groomed, always attired in a freshly pressed suit, with his thick, dark hair brushed to perfection. Edison had no affinity for learning foreign languages. Tesla spoke several languages fluently. Edison was not exceptionally good at mathematics and regarded it as a mere tool. For Tesla, mathematics was the highest calling, and working on mathematical problems gave him nothing but the greatest pleasure.

Lacking Edison's support, Tesla tried to start his own company, but it soon failed. For a while he wandered the streets of New York poor and alone, at one point working as a ditch digger to earn a living, a sore trial for the fastidious scholar. He didn't surrender his goals, though. He started another company with two backers and in 1888 presented a paper on alternating current at the American Institute of Electrical Engineers. Among the members of the audience was George C. Westinghouse.

The Rise of AC

George C. Westinghouse had made his fortune by inventing an air brake for locomotives in 1869. Once his air brake manufacturing company became a success, he expanded into electricity. Westinghouse was one of the few engineers who saw the value of AC over DC. He had developed an AC generator of his own and achieved some success wiring part of his home city of Pittsburgh for electric light. Edison regarded him as a bitter foe. Westinghouse recognized Tesla's generator as far superior to his and offered to buy Tesla's company. Westinghouse was a shrewd businessman and got the far better deal, for he bought all of Tesla's AC patents along with the company. But Tesla had acquired a secure financial footing for the first time in his life and, more important, a chance to demonstrate the full effectiveness of AC power.

Westinghouse and Tesla scored a major coup when they were asked to illuminate Chicago's World Columbian Exposition in 1893. Over 180,000 lightbulbs powered by AC lit the fairgrounds, turning it into the "White City." The press and public lauded the two men as heroes of a new, modern electric age. In 1896, Westinghouse installed the world's first hydroelectric generating plant at Niagara Falls. AC ruled and would continue to do so as electricity spread across the globe.

Yet no one knew exactly why alternating current was so much more effective than direct current. Even Tesla, with all his brilliance, could not deduce the activity or mechanism within electricity that gave AC its edge. The answer to that riddle lay in the hands of the English physicist J. J. Thompson.

While Westinghouse and Tesla were lighting up the World's Fair, Thompson was quietly demolishing a belief held for centuries. Atoms were not indivisible. They could be broken down into even smaller particles. One of those particles he called an electron.

British physicist J. J. Thompson verified the existence of the electron. His observations contributed to the development of atomic energy.

THOMPSON DISCOVERS the ELUSIVE ELECTRON

A professor of physics at Cambridge University, J. J. Thompson was particularly interested in cathode light rays. A cathode ray tube is a glass vacuum tube with two metal wires inserted at either end to create a positive and negative anode. When the wires are charged with electricity, a beam of light passes through the tube from one end to the other. Researchers had noticed that the beams passing through this vacuum always moved in a straight line rather than radiating outward like that of an electric lightbulb. Thompson knew that light traveled in waves, but the beam in the cathode tube didn't behave as if it were made of waves; it behaved as if it were made of particles. Particles, however tiny, had mass, and mass, he knew, could be measured and weighed.

Thompson used Maxwell's equations to devise a series of complex experiments on cathode rays. Essentially, he used magnetic and electric plates to deflect the rays. By this he established that the rays were made up of negatively charged particles because they were deflected by a positively charged plate. Particles so small could not be measured directly, but Thompson used the size and angle of the deflection plus the amount of heat generated by the particles to come up with an approximate weight. His research revealed that these particles were far lighter than even hydrogen, the lightest of all atoms. Atoms, he concluded, consisted of smaller particles with positive, negative, or neutral charges. The negatively charged particles were electrons. Electricity was generated by the movement of electrons from one atom to another.

How did this help explain the difference between AC and DC? To put it very, very simply, AC generates energy by "jostling" electrons back and forth between atoms. AC generators use lower current than DC generators. This means the electrons encounter less resistance than they would with DC.

The higher current used by DC generates more heat, causing the cables carrying the current to become hot. As they do so, some of the electric energy generated by the electrons is absorbed by the insulating materials around the cables and lost or "thrown away." Although it seems counterintuitive, the higher the current the more likely it is to lose power over a long distance.

Thompson published his findings in 1899. The discovery of the electron did far more than explain the nature of electricity. It was a major breakthrough in physics and opened the door to the development of nuclear fission and atomic energy in the following century. On another, more intimate level, the electron would help reveal secrets of biology deep within the heart, nerves, and brain of every human being.

The White City

The World's Columbian Exposition, or Chicago World's Fair, as it was also known, was held in 1893 to commemorate the four-hundredth anniversary of Christopher Columbus's first voyage to the Americas. During the year it was open, over 27,300,000 people visited the fair, which covered 690 acres (279.2 hectares) near the shores of Lake Michigan. The fair's forty-six main buildings were painted to look like pure white marble, an effect enhanced by the glow of electric streetlights at night.

An entire building was devoted to electrical exhibits, including a gigantic lighthouse light and Edison's Electric Column, a metal tower covered by five thousand individual lightbulbs. The fair's guidebook called the exhibit, "a museum of enchantments where the eye of curiosity was almost blinded by bewilderment and dazzled by surprises."

The fair also mounted the world's first Ferris wheel. Standing at 137 feet (41.75 m) high, it had thirty-six cars, each of which carried sixty passengers. A ride was two revolutions and cost twenty cents.

Perhaps the most amazing exhibit of all, however, was in the Egyptian Pavilion where young women performed the *danse du ventre,* or belly dance. One dancer gyrated with such "wild abandon" that scandalized fair officials felt it necessary to call the police.

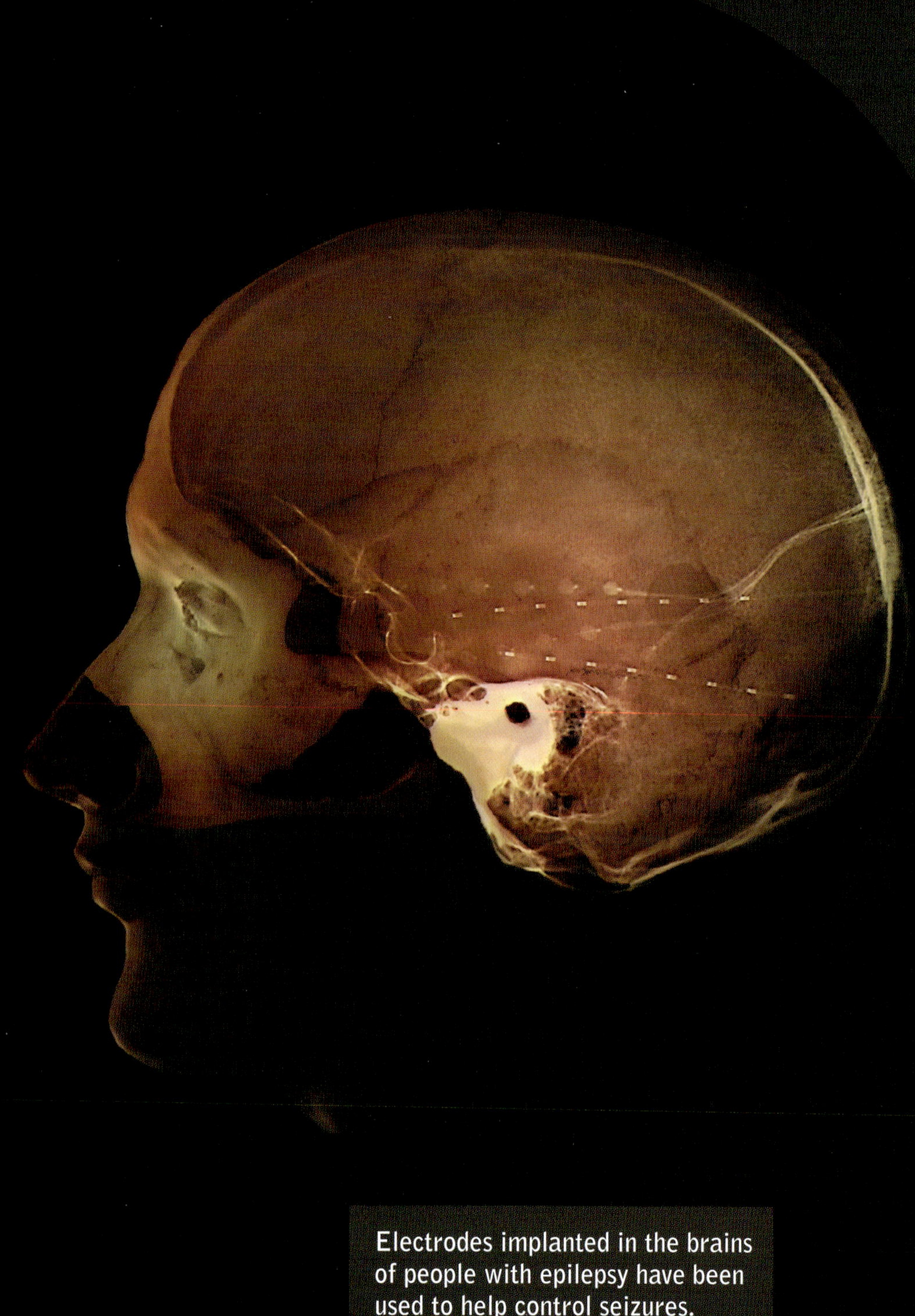

Electrodes implanted in the brains of people with epilepsy have been used to help control seizures.

CHAPTER 5

Electric Medicine: A New Age of Electricity

One afternoon in 2006, Mathew Nagle sat down at his computer and did the things a lot of other people do. He opened his email and then played a video game for a few minutes. Later, he watched a bit of television, switching the channel back and forth. There was nothing unusual in this activity, except for the fact that Nagle never touched a computer key, mouse, or remote control device. Nor did he use voice commands. Nagle operated his computer and television completely with his mind. A few years earlier, the twenty-six-year-old Nagle had become paralyzed from the neck down. He could not move his arms or legs. He could pilot his wheelchair via a sip and puff tube, but the respirator that helped him breathe made it difficult for him to speak. Instead he relied on a **BCI**, or brain-computer interface, in this case a set of tiny microelectrodes implanted deep within his brain, to move the cursor on his computer screen.

Nagle's BCI had been developed by Tom Donoghue, a professor of neuroscience at Brown University. Donoghue called the system Braingate. The electrodes in Nagle's brain were attached to an external connector plate embedded in Nagle's scalp. When he was ready to use his computer, an assistant would run wires from the plate to an electronic device that, in turn, was connected to the computer. Over time, Nagle and Donoghue's team had trained the computer to read Nagle's brain signals. When Nagle thought of moving the cursor to the right, for instance, the neurons in his brain would emit a particular pattern of signals that the electrodes picked up. While Nagle imagined his action, the researcher would move the cursor right. When Nagle pictured himself clicking on a certain icon, the process would be repeated. Eventually, the computer "learned" which patterns of signals were related to each action and could "read" Nagle's mind. The computer was responding to electric signals generated by Nagle's thoughts.

After a few months, Nagle could execute several basic computer commands. He even progressed to the point of playing a simple computerized Ping-Pong game with an able-bodied opponent. Nagle died at the age of twenty-seven, three years after receiving his BCI implant. Dr. Jon Mukand, who had worked with Nagle, praised him as "a brave pioneer." "He wanted to try to do everything possible to take advantage of the latest research," Mukand told reporters, "not only for himself, but for everyone else with severe disabilities." In the years following Nagle's death, BCI's would undergo even greater development, enabling paralyzed individuals to not only move computer cursors, but also to control robotic arms skillfully enough to pour a cup of coffee and complete other mundane tasks so necessary to everyday life. One of the oldest uses of electricity, it seems, is also the newest. Ancient people often believed that electricity had healing properties. Modern science has proven them right, though in ways they could have never predicted.

WITH EVERY BEAT of YOUR HEART: ELECTRO-CARDIOLOGY

During the late eighteenth century, Luigi Galvani's experiments with frogs had led him to hypothesize that electrical energy might be inherent in all animals. This notion was dismissed

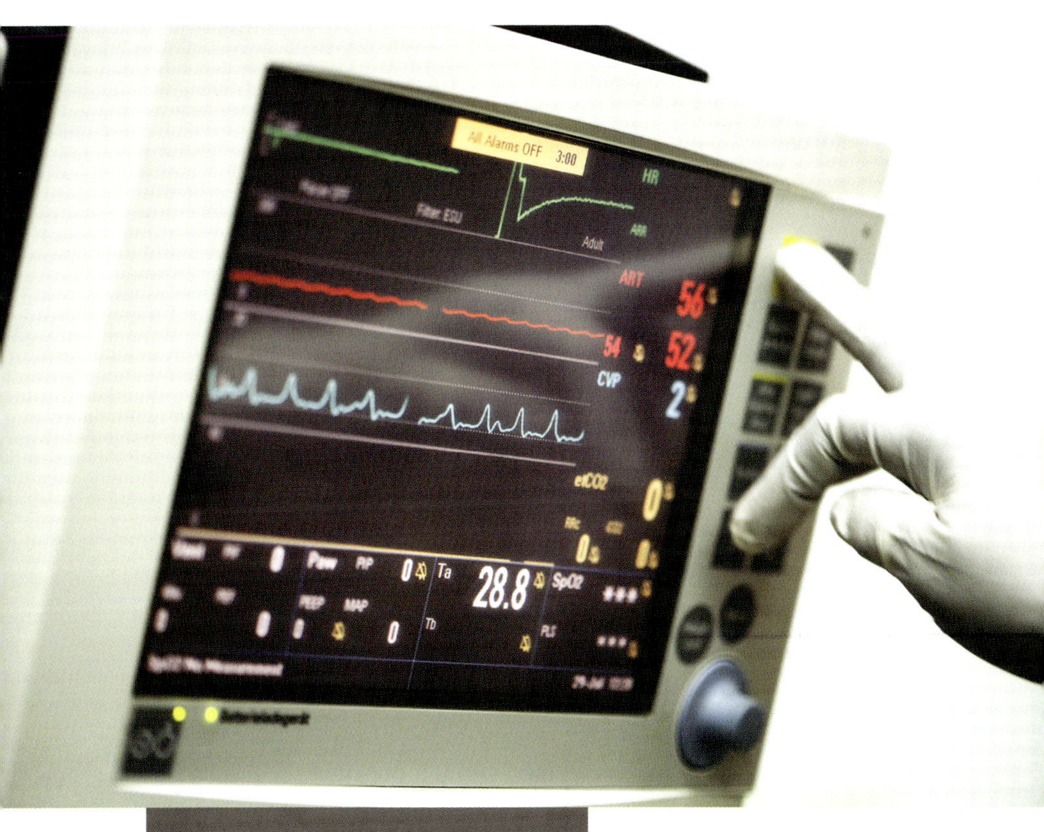

The electrocardiogram gives physicians a visual graph and record of the electrical activity in a patient's heart.

by his contemporaries, but it was not completely forgotten. In 1901, slightly over a century after Galvani died, the Dutch physiologist Willem Einthovan developed a machine he called the string galvanometer, a device that used wire and magnets to detect the electrical impulses released by a beating heart.

Less than a decade earlier, in 1893, Wilhelm His Jr., a Swiss anatomist, had discovered and described a special type of muscle tissue within the heart that transmitted electrical impulses and synchronized the heartbeat. Einthovan's string galvanometer could detect irregularities in cardiac rhythm, which helped physicians diagnose heart disease. Einthovan won the Nobel Prize in Medicine in 1924, and his galvanometer became known as the electrocardiograph. By the end of the twentieth century, electrocardiograph machines had become common in hospitals. In the early twenty-first century, electrocardiograph machines moved out of hospitals and into the field with miniaturized versions that could be attached to the patient's fingertip and transmit information wirelessly. In 2016, Korean researchers streamlined the machine even further, developing an electrocardiograph skin patch that could send out signals even underwater. The texture of the patch was modeled on the footpads of the wall-climbing gecko. *Chemical & Engineering News* hailed it as "crucially important in next-generation skin-like technologies for wearable electronics."

So-called wearable electronics can be worn outside the body or within it. One of the oldest and most well known is the artificial cardiac **pacemaker**. Early pacemakers were external devices used only in emergencies. Dr. Mark C. Lidwell of Sydney, Australia, invented the first successful device using a needle attached to an electrical outlet. By plunging the needle repeatedly into the arrested heart muscles, he could reanimate a dying patient. In 1926, he revived a stillborn infant whose heart had stopped and who was, for all practical purposes, dead until Lidwell and his pacemaker intervened. Instead of publicizing this event as a heroic rescue, Lidwell chose to keep it from the popular press. When reports did leak out, his name did not appear in any of the news articles. Readers knew only that an unnamed doctor had resuscitated an infant using an electric machine. Bringing the dead back to life was not something people took lightly, even in the 1920s. Lidwell legitimately feared he would be accused of assuming ungodly powers over his patients. Though it may seem ridiculous in retrospect, the shadow of Mary Shelley's *Frankenstein* still loomed over this type of cardiac research. Any doctor suspected of using electricity to revive the dead might be deemed more a mad scientist than a savior of humanity. While his medical colleagues knew of his work, Lidwell's role in the development of the pacemaker only reached the public after his death at the age of ninety in 1969.

Pacemakers, Evolved

By the time Lidwell died, pacemakers had saved millions of lives. Rather than needles, they relied on batteries that could stimulate the heart continuously. The first battery-powered pacemakers were strapped to patients' chests and delivered electric impulses through the skin. In 1958, Swedish doctors implanted a small pacemaker directly into the chest cavity of forty-three-year-old Arne Larsson. Over the next forty-three years, Larsson would receive a total of twenty-six pacemakers, an average of one every twenty months. The constant operations must have been grueling, but Larsson was a hardy patient, and he survived to the age of eighty-six, outliving both of the surgeons who had performed the original implant.

The work of American engineer Wilson Greatbatch also refined the pacemaker by increasing its life and reducing its size. Greatbatch was not a physician but an inventor and tinkerer in the great tradition of Edison. During his lifetime, he would receive 325 patents for inventions ranging from methods to diagnose AIDS to a solar-powered paddle-free canoe. A former radio operator for the US Navy during World War II, he had majored in electrical engineering at Cornell University. A part-time job in the animal lab at Cornell introduced him to some cardiology researchers. Later, while working at Buffalo State University in the 1970s, he produced a pacemaker less than 2 square inches (12.9 square centimeters) that was fueled by a lithium battery. Previous pacemakers had used less reliable zinc batteries. The lithium pacemaker could last for up to six years, cutting down on the frequent and often risky operations. Though Greatbatch's design was not employed in the first implant in Sweden, it was in widespread use by the 1980s. By the time Greatbatch died in 2011, over a half million lithium-powered pacemakers were being implanted in patients every year.

The next breakthrough in pacemakers came with the advent of **nanotechnology** in the early twenty-first century. In 2009, Angela Belcher, an engineer at the Massachusetts Institute of Technology, used genetically engineered viruses to pull lithium from a saline solution, creating a biological battery capable of powering a small pacemaker. In 2013, researchers used her bio-battery to develop the prototype for a pacemaker that was no larger than a pill and could be threaded into the heart with a catheter.

The heart was not the only organ to benefit from advances in electrophysiology. While Einthovan, Lidwell, and Greatbatch devoted themselves to cardiology, other researchers had begun to investigate the electrical impulses in the most mysterious organ of all—the human brain.

A "SHOCK in the BRAIN": The FALL and REVIVAL of ELECTROCONVULSIVE THERAPY

Alessandro Volta experienced a sudden fright when he placed the electrodes of a battery in his ears. "The disagreeable sensation," he wrote, "… which I apprehended might be

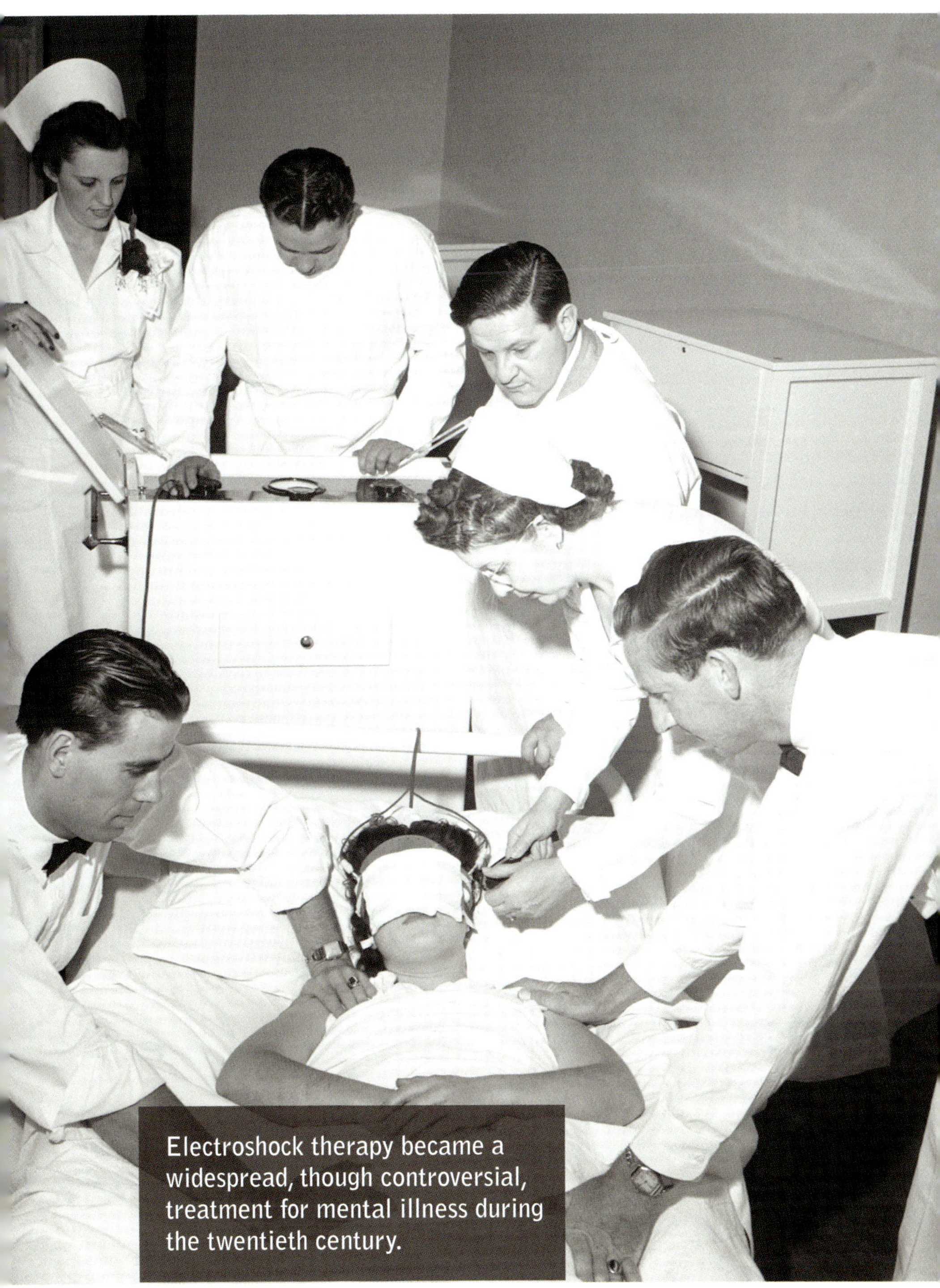

Electroshock therapy became a widespread, though controversial, treatment for mental illness during the twentieth century.

dangerous, of a shock in the brain, prevented me from repeating the experiment." Volta was reflecting upon his invention of the electric battery in 1800. He had been curious about the effects of electricity on various parts of his body, but that curiosity stopped at his brain. A shock to the inner sanctum of the skull was not something he wanted to feel twice.

Volta was not the first to notice the effects of electricity upon the mind. The ancient Roman physician Scribonius Largus recommended the use of shocks from a "torpedo ray" to cure headaches and also relieve pain from gout in a list of various medical treatments he drew up in 47 CE. Other early people may have used electric fish in rituals to banish evil spirits. A sixteenth-century Jesuit missionary in Ethiopia noted that local tribesmen applied electric catfish to the bodies of individuals thought to be possessed by the devil, though he did not record the results. In the early eighteenth century, Michael Schuppach of Switzerland became known as the "Miracle Doctor" when he cured a farmer who insisted that he was possessed by eight devils. Reports in a contemporary medical journal indicate that the doctor approached his patient with tongue firmly in cheek. To set up his cure, he hid a static electricity machine behind a screen and told the farmer to grasp a copper wire firmly in his hand. The doctor then charged the machine. "Presently there was a loud shriek … 'That,' remarked the doctor, has gotten rid of your first devil." The treatments continued for eight days. On the eighth and last day, the doctor told his patient that the remaining spirit was none other than Beelzebub, the prince of devils. "Thereupon he administered a shock so severe, it threw his patient on the floor. A complete cure resulted."

Throughout the eighteenth and nineteenth centuries, electric shocks were used to cure a variety of mental ailments. The French physician Desbois de Rochefort reported using electricity in the 1770s to cure despondency caused by deep grief. He also warned of the "dangerous excitement" that might result from too many applications of electric shocks to the head. In 1760, an article in *Gentleman's Magazine* described how the British doctor William St. Clare had treated a group of young women working in a London textile factory who had been seized by mass hysteria. It began when one girl tossed a mouse at another, who, having a phobic reaction, collapsed in convulsions. Within a few weeks, dozens of girls were experiencing "anxiety, strangulation, and very strong convulsions." So severe were these symptoms that some of them began tearing their hair out and "dashing their heads against the floor or wall." The doctor arrived at the factory with his portable electric machine, and "by electric shocks, the patients were universally relieved, without exception." In 1853, a medical journal noted that, "No nervous affliction whatever should be regarded as incurable until some form of electricity has been tried."

Yet the effects of electricity were mostly anecdotal and there was no systematic research on shock therapy. The great psychiatrist Sigmund Freud regarded electricity as similar to hypnotism and dismissed it as treatment that worked by suggestion and was therefore no more scientific than magic or astrology. Electric shock therapy did not enter modern medicine until 1937 when the Italian neurologist Ugo Cerletti announced that he had used shocks to cure patients diagnosed with schizophrenia and other serious forms of mental illness. His method worked by applying an electric current powerful enough to cause a series of controlled seizures. Cerletti believe these seizures had a positive effect on the brain, calming agitation and allowing the patient to behave normally afterward. Cerletti had engaged in extensive **clinical trials** to support his hypothesis, and his work was greeted with enthusiasm by the scientific community. Metal illness was extremely hard to treat, and electroconvulsive therapy, or **ECT**, seemed to give sufferers and physicians genuine hope. The use of ECT spread throughout Europe and the United States, and it rapidly became a standard treatment in mental institutions.

The Dark Side of ECT

Unfortunately, ECT was often abused. Some doctors and nurses simply did not know how to control the machine and delivered enormous shocks to patients without intending to. In other cases, intentions may have been far more sinister. Critics claimed ECT was employed to punish patients who disobeyed the rules or who questioned authority. By the 1970s, ECT was regarded as a weapon rather than as a treatment. In her autobiographical novel, *The Bell Jar*, American poet Sylvia Plath recounted her first experience with ECT when she was being treated for depression in the 1950s:

> I shut my eyes. There was a brief silence, like an indrawn breath. Then something bent down and took hold of me and shook me like the end of the world. Whee-ee-ee-ee-ee, it shrilled, through an air crackling with blue light, and with each flash a great jolt drubbed me till I thought my bones would break and the sap fly out of me like a split plant. I wondered what terrible thing it was that I had done.

Public outrage over the excessive use of ECT caused the practice to decline dramatically in the late twentieth century, and ECT became regarded as something left over from the brutal dark ages of psychiatry when patients had little dignity and few rights. The issue, however, was the abuse of ECT, not its use. A statement released by the National Institutes of Health in 1985 acknowledged that while ECT often had serious side effects, it had still been proven effective in cases of severe depression that could not be alleviated by medication.

Gradually, modifications in ECT helped it become accepted once more, at least by some patients and practitioners.

ECT Makes a Comeback

"There's been a sea change in our understanding of ECT in the last decade," Dr. Harold Sackheim of the New York Psychiatric Institute told a reporter for the *New York Times* in 2008. He added that electric currents could now be adjusted based on the size and condition of the patient. Treatments involved smaller doses of electricity given in shorter durations. New knowledge of how the brain worked also helped. ECT was notorious for causing memory loss. One innovation was to apply the current only to the right side of the head so it would have less of an effect on the areas of the left side of the brain that govern verbal processing. Though the treatment remains controversial, it is gaining greater acceptance.

Perhaps one reason for that acceptance is the increase in our knowledge of how electricity works within our bodies, carrying messages back and forth between our vast and complex network nerves and our solitary brain.

The BODY ELECTRIC: ELECTROPHYSIOLOGY and the NERVOUS SYSTEM

When American poet Walt Whitman wrote, "I sing the body electric" in 1855, he was using metaphor. On another level, however, he was right. Electricity courses through every nerve of the human body, sending its signals ceaselessly to every nerve and muscle fiber. Only a few years before Whitman had referred to electricity in his poem, Hermann von Helmholtz, a physiology professor, mathematician, and philosopher at the

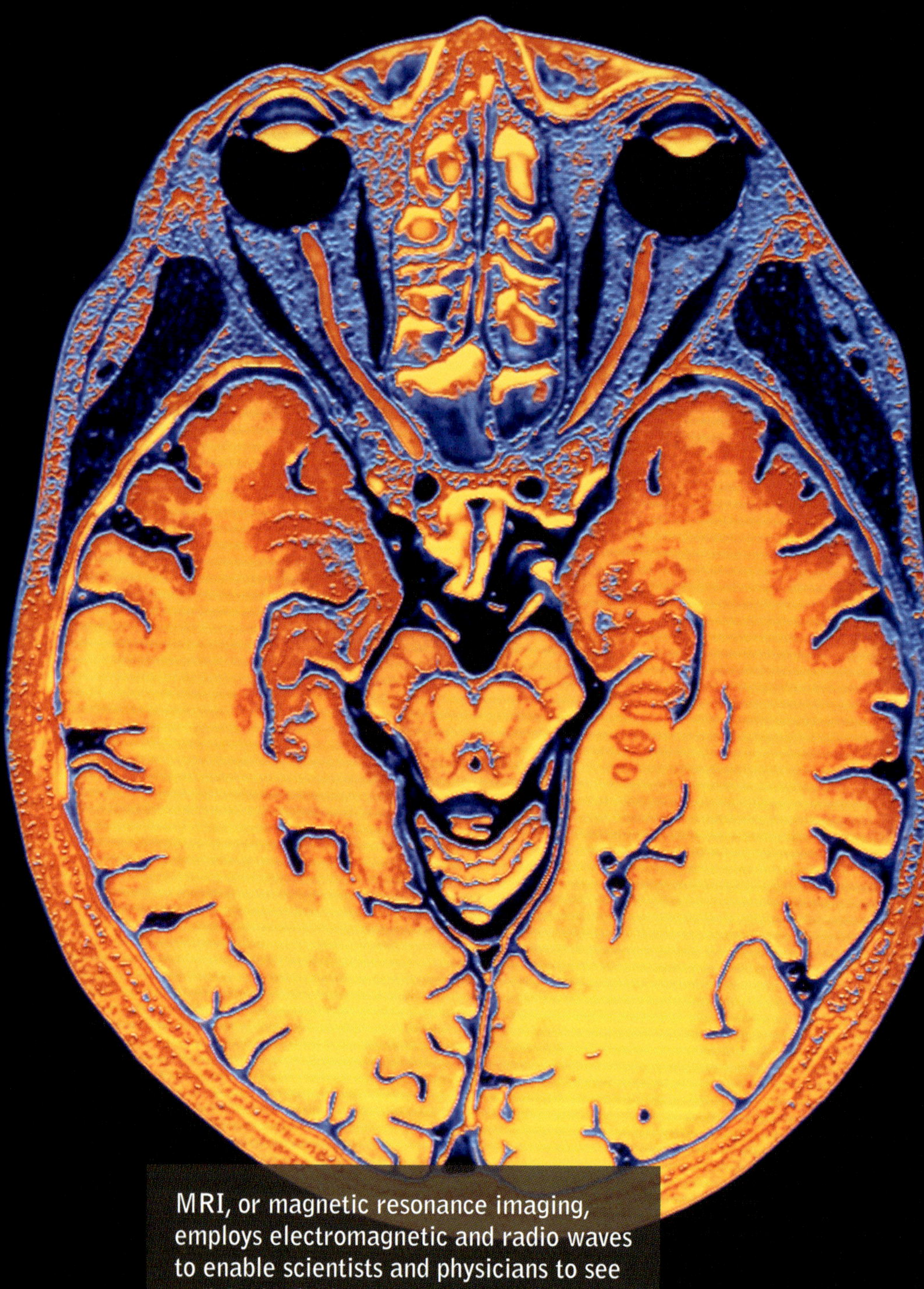

MRI, or magnetic resonance imaging, employs electromagnetic and radio waves to enable scientists and physicians to see activity in the brain.

University of Konigsberg, Germany, had measured the speed at which this biological electricity traveled.

Helmholtz was a scientist of so many achievements that this particular observation was almost like a footnote to his long career. He is most known for his theories on the conservation of energy and his extensive work in electrodynamics and optics. He also dabbled in acoustics, astronomy, and geology, and invented the first ophthalmoscope, which allowed physicians to examine the inside of the human eye. Helmholtz had started his research as a student of medicine before pursuing an advanced degree in physiology and physics. He was no doubt aware of Galvani's earlier work with frogs and had developed his dissertation on the physiology of the nervous system.

In 1849, he devised a method for measuring the electric current within a nerve. He attached the two electrodes of a galvanometer to each end of the exposed nerve in a frog's leg and noted the length of time it took for an electric impulse to travel from one terminal to the other. Like many other researchers, Helmholtz assumed that electricity would travel through a nerve almost instantaneously, at the speed of light. Scientists at that time were well aware of the telegraph and knew that electricians had already measured the rate at which electricity traveled through the telegraph wires as fairly close to that of light, around 186,000 miles per hour (299,792 kilometers per hour). He was very surprised, therefore, to discover electricity coursing through the frog's leg at a mere 100 mph (160.9 kmh). Of course, this is still very fast, given the size of a frog or even a larger creature, such as a human being. But it was far slower than Helmholtz had expected. On one hand, the slow speed made sense, for if the current traveled at the speed of light, the nerve would become so hot it would burn up. Helmholtz, though, had no idea what made electricity in the body so much slower than when it went through metal conductors. The inside of the body was liquid, after all, and water was a very good conductor of electricity indeed.

Helmholtz soon put this problem aside as he moved on to other areas of research. He knew that many scientific discoveries seemed to have no particular value when they were made, an idea he expressed in an often quoted sentence from his essay, *Academic Discourse*, published in 1862: "Whoever in the pursuit of science, seeks after immediate practical utility may rest assured that he seeks in vain."

Helmholtz's discovery would prove to have practical value but only well after J. J. Thompson's discovery of the electron in 1899 had helped scientists understand how electricity actually worked. Electrons in a living body, however, seemed to work somewhat differently than those in a metal wire, and scientists still did not know why. So, in 1939, Alan Hodgkin went out looking for a giant squid.

Hodgkin's Squid

Hodgkin had become interested in how impulses traveled along the nerves while studying physics and chemistry at Trinity University. Hodgkin believed that the electric currents running through the nerves might be related to the presence of sodium **ions** in the body. An ion is an atom with an irregular number of electrons. Sodium is an element that gains or loses electrons very easily. Nerves in most animals are too small for the action of sodium ions to be observed closely, but the giant squid has singularly large nerves, not just because it is a large animal, but because its tentacles need to whip out and snap back with extraordinary speed when it seizes its prey. Hodgkin assumed, correctly, that if sodium ions were instrumental in conducting electrical impulses along the nerve, the squid would have a lot of these ions clustered along its nerves. He had seen some initial experiments using squids during a visit to the

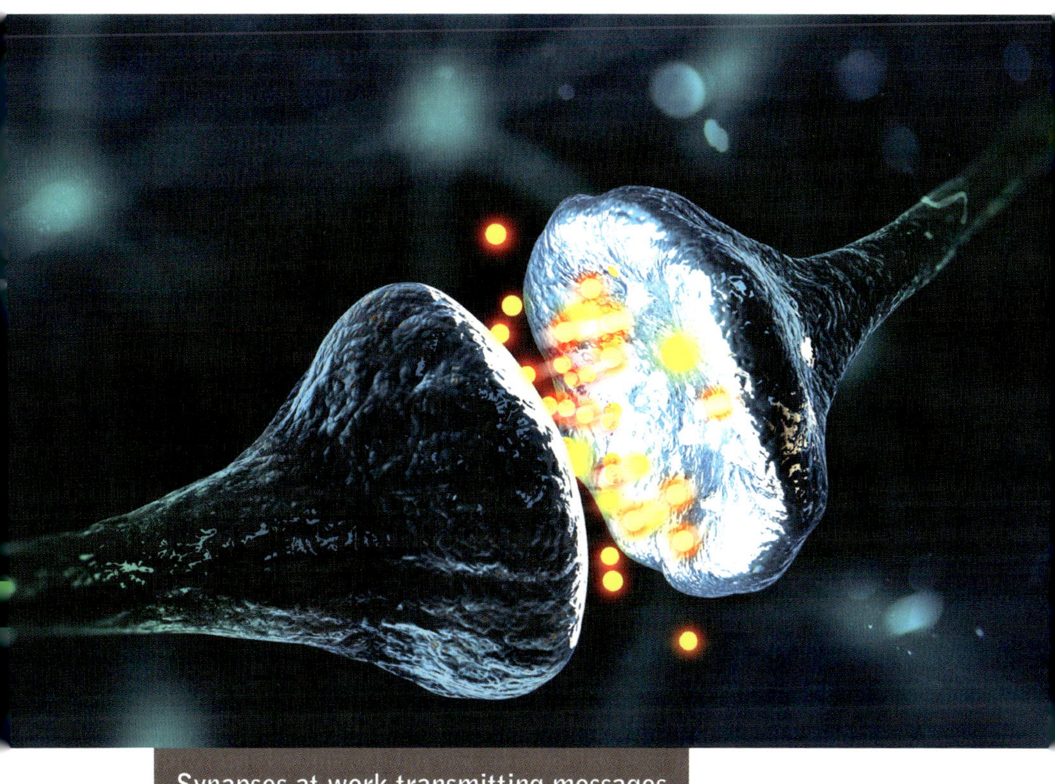

Synapses at work transmitting messages

Woods Hole Laboratory in Massachusetts the year before and wanted to build on what he had learned there.

Hodgkin enlisted the help of a younger colleague, Andrew Huxley. In addition to being an excellent physiologist, Huxley was also a gifted mechanic. He had built his own internal combustion engine as a boy and could construct any kind of equipment Hodgkin needed. Finding a good-sized squid back home in Plymouth, England, proved difficult, though. Together Hodgkin and Huxley spent most of the summer of 1939 scouring fish markets without success. Hodgkin even went out on the fishing trawlers himself, again without luck. A large squid was not a common catch. The fishermen must have sympathized with the scientist, though, for when he returned after a brief vacation, he found a gift in the form of a pile of humongous, glistening squid. The fishermen told him that the best season for hunting squid off the coast of Britain was in the fall. He remembered that when he returned to his work after World War II and made sure he always spent his autumns in a laboratory near the fishing docks of Plymouth where specimens would be plentiful.

Hodgkin and Huxley made some initial investigations into nerve impulses, but they were forced to abandon their research during the war. Both were recruited to work on the development of radar. Though they served in different units, Hodgkin would often contact Huxley when his unit needed a piece of original equipment. Huxley would then get out his tools, go to work, and the equipment would be ready within a few days. What they learned about electricity while working with radar later helped them when they went back to their physiology experiments in 1945.

The sodium ions, they realized, might be acting as "boosters," helping the electrical current move along the nerve. By stimulating the large nerve thread, or axon, of a squid, Hodgkin and Huxley could observe the activity of the sodium ions as they poured through the certain gaps or "gates" that opened in the wall of the axon. As they filled the axon, these sodium ions would begin dropping and picking up electrons to send the signal along. After the signal reached the end of the nerve, the nerve would then rid itself of all the sodium ions and rebuild the wall of the axon. Later, Hodgkin and Huxley discovered that potassium ions in the body also worked in conjunction with sodium ions to convey nerve impulses to and from the brain. Because potassium and sodium ions are large they move more slowly than the smaller electrons speeding through a metal wire.

This research provided the basic explanation for how nerves transmit the electrical signals that control every function of the body, both voluntary and involuntary, from running a race to rapid eye movements in our sleep. Hodgkin and Huxley received the Nobel Prize in Medicine or Physiology for their discoveries in 1963, along with John Eccles of Australia. Eccles had enlarged upon the research of Hodgkin and Huxley by demonstrating how the electrical impulses passed between the gaps, or **synapses**, between nerves via chemicals known as neurotransmitters. The accomplishments of these three scientists created the modern discipline of electrophysiology, the study of electricity's role in all the body's physiological functions.

Hodgkin's and Huxley's Legacy

In the later decades of the twentieth century, scientists would develop over twenty different branches and sub-branches of electrophysiology, among them: EPG, electropneumography, the study of lungs; EOG, electrooculography, the study of the eye; EGG, electrogastrography, the study of the stomach; and EEG, electroencephalography, the study of the brain.

EEG has led to the development of **neurofeedback**, which allows people to see and monitor their own brain activity via electrodes attached to the skull. When the electrodes are stimulated, an EEG machine produces a printout showing brain activity as a series of waves. Neurofeedback, also known as biofeedback, has been used to help people with conditions such as ADHD (attention-deficit/hyperactivity disorder) by giving them a tool to measure their own mental states. They can literally "see" when they are becoming tense, anxious, or hyperactive and use deep breathing or other relaxation techniques to regain control of their emotions. In addition, neurofeedback is increasingly used outside of hospitals and clinics to enhance general well-being and performance. Athletes, dancers, and artists have begun to employ various biofeedback machines to help them identify and attain states of peak awareness, physical perception, and creativity.

Minds and Machines

People have always tried to come up with an explanation for how the body works. In the eighteenth century, the human body was seen as an intricate piece of clockwork, something that was wound up at birth and whose inner workings operated with mechanical precision. In the nineteenth and twentieth centuries, the body was envisioned as an elaborate telegraph or telephone system, with the brain serving as a switchboard sending and receiving the electric signals that coursed along the nerves. In the twenty-first century, the brain has been described as a kind of biological computer, endlessly processing, storing, cataloguing, updating, and modifying the vast amounts of data that flow to us from every cell of our bodies.

Yet none of these metaphors quite fit. We are no more living computers than we are living clocks. The study of electrophysiology has demonstrated that we are all in a state of constant flux and change. Every electrical impulse in our nerves creates balance and imbalance. Ions, electrons, and neurotransmitters all interact in intricate patterns we have only begun to understand. Over the past centuries electricity has changed how we live and work. Now it has begun to change what we know about our own minds and how we see ourselves.

LOOKING AHEAD

Of course, scientists today look to revolutionize electricity's applications beyond health and medicine. Finding alternative sources of power is a mission that grips some of our most innovative thinkers. This is a lofty goal, and much is at stake, including Earth's climate. Though the mystery of electricity traces back to ancient times, and our modern understanding is sophisticated, there is still work to be done.

Alfred Nobel and His Dynamite Prizes

In 1888, Swedish chemist and businessman Alfred Nobel opened a newspaper and learned he was dead. What's more, he wasn't even mourned. "The merchant of death is dead," crowed the reporter. The obituary was really for his brother Ludvig who had died a few days earlier. The mistake was easily corrected, but the words still stung. Nobel considered himself a man of peace. In 1867, he had come up with a method for stabilizing the highly volatile explosive nitroglycerin and molding it into sticks. He called this dynamite. Dynamite made nitroglycerine far safer and easier to handle. After switching to dynamite, workers in his many weapons factories suffered fewer injuries and accidental deaths. By the late 1880s, Nobel had amassed a fortune.

Nevertheless, he was disturbed by the idea that he might go down in history as a man who got rich profiting from human slaughter. Nobel didn't stop manufacturing weapons, but he did draw up a will leaving his entire fortune to a foundation that would distribute annual prizes to "those who, during the preceding year, should have conferred the greatest benefit to mankind." The categories for the prizes were literature, chemistry, physics, physiology or medicine, and peace. Another prize for economics was added in 1968. Between the establishment of the foundation in 1901 and 2016, 911 prized have been awarded, thirty-six of them for scientific achievements related to electricity.

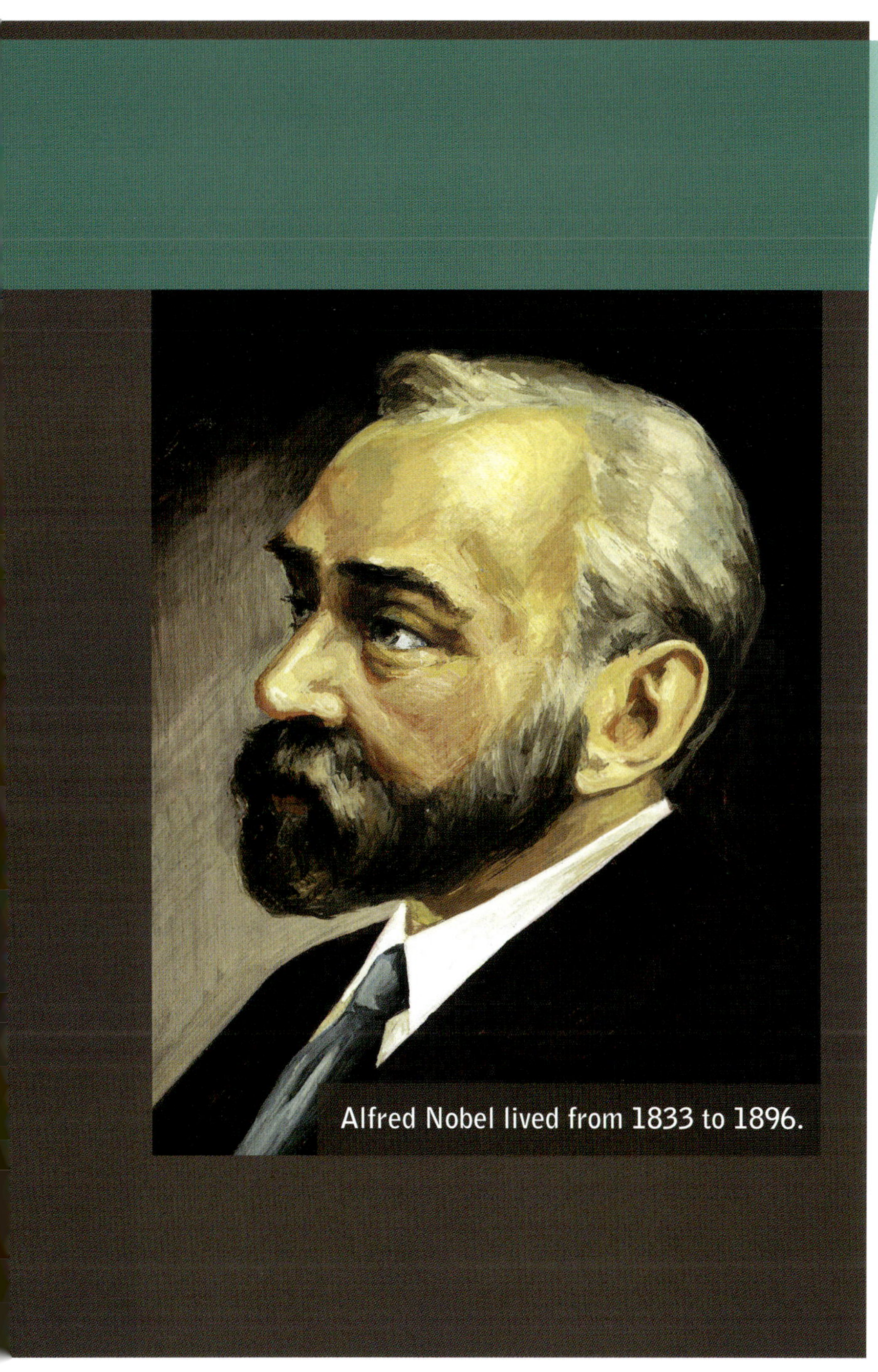

Alfred Nobel lived from 1833 to 1896.

Chronology

600 BCE Thales of Miletus writes first description of static electricity

400 BCE Democritus describes the world as consisting of indivisible particles he calls atoms

47 CE Scribonius Largus describes using electric fish to cure headaches

1640s Pierre Gassendi hypothesizes that matter consists of atoms joined together to form molecules

1730 Stephen Grey conducts electricity along a single thread for 700 feet (213.3 m)

1740s Pieter van Musschenbroek invents the Leyden jar

1747 Benjamin Franklin develops the lightning rod

1750 Franklin describes electricity as consisting of particles

1752 Thomas-François Dalibard and George-Louis Leclerc demonstrate that lightning is electricity

1770s Desbois de Rochefort uses electric shocks to treat depression

1791 Luigi Galvani publishes his research on electrical impulses in the nerves of frogs

1800 Alessandro Volta invents the battery

1808 John Dalton develops Dalton's law

1820 Hans Christian Øersted publishes his research on electricity and magnetism

1825 Andre Marie Ampere develops Ampere's law

1827 George Simon Ohm develops Ohm's law

1831 Michael Faraday discovers electromagnetism

1844 Samuel Morse demonstrates telegraph using Morse code

1849 Hermann von Helmholtz measures the speed at which electricity travels through a nerve

1857 James Maxwell develops Maxwell's equations

1861 Transcontinental telegraph completed in the United States

1870 First transatlantic telegraph cable laid between North America and Great Britain

1880 Thomas Alva Edison receives US patent for the lightbulb

1893 George Westinghouse and Nicola Tesla demonstrate AC electric lighting at Chicago World's Fair

1899 J. J. Thompson verifies the existence of the electron

1901 Willem Einthovan invents first electrocardiograph

1924 Mark C. Lidwell invents first pacemaker

1937 Ugo Cerletti develops modern electroconvulsive therapy (ECT)

1947 Alan Hodgkin and Andrew Huxley discover how electrical impulses pass through the nerves

1958 First internal pacemaker implanted in a patient

1970s Wilson Greatbatch develops pacemaker powered by lithium battery

2000s First brain-computer interfaces (BCI) developed

2009 Angela Belcher develops a battery using genetically engineered viruses

Glossary

AC An electric current that reverses direction at frequent intervals.

amp Also known as an ampere, an international unit of electric current.

anode A positively charged electrode.

atomism The theory that something can be understood by breaking it down into its basic, elementary components.

BCI Brain-computer interface, a electric device that enables the brain to connect directly with a computer.

carbonize To convert something into carbon by heating or burning it.

cathode A negatively charged electrode.

clinical trial A research study that compares the effects of a particular treatment by using different groups of patients.

DC An electric current that flows in only one direction.

dissection The process of cutting open a plant or animal to study its inner workings.

dynamo A machine that transforms mechanical energy into electrical energy.

ECT Electroconvulsive therapy, a form of therapy that uses electrical currents to induce convulsions.

electrocardiograph A machine that measures and records the activity of the heart and produces a visual graph of that activity.

electrochemistry The study of the reactions between chemicals and electrical forces.

electrode A wire, node, or other conductor through which electricity flows into or out of a battery.

electromagnetism The interaction between electric currents or fields and magnetic fields.

electron A particle that forms part of an atom and is charged with negative electricity.

electroplating The process of adding a very thin coat of metal to another metal by means of an electrical reaction.

element A chemical that forms one of the primary components of matter and cannot be broken down into simpler substances.

galvanize To energize or compel to take action.

generator A machine that generates energy.

humanism A philosophy the emphasizes the study of human thoughts and feelings over the study of theology or spiritual matters.

induction The process of creating an electric current by the motion of a magnet.

ion An atom with a net positive or negative charge caused by the gain or loss of an electron.

Leyden jar A glass jar charged with static electricity and lined with metal foil.

lightning rod A metal rod attached to a structure to divert electric charges from lightning to the ground.

mechanical engineering A field of engineering that deals with the construction and repair of machines.

metaphysics The study of things that are beyond human experience.

methane An odorless, colorless gas that is the main component in natural gas.

nanotechnology Technology that deals with the manipulation of atoms and molecules.

nervous system The network of nerves and fibers in a body that transmits impulses to muscles and organs.

neurofeedback A computer-aided process for helping people monitor and control their own brain activity.

oxidation A chemical process by which electrons are lost.

pacemaker A device for regulating the contractions of the heart muscles.

spiritualism A belief that one can communicate with the dead.

static electricity An electric charge produced by friction.

synapse A gap between two nerve cells.

transformer A device that increases the voltage of alternating current.

Further Information

BOOKS

Bodanis, David. *Electric Universe.* New York, NY: Crown Publishers, 2005.

Casarett, David. *Shocked: Adventures in Bringing the Recently Dead Back to Life*. New York, NY: Current, 2014.

Freeberg, Ernest. *The Age of Edison: Electric Light and the Invention of Modern America.* New York, NY: Penguin Press, 2013.

James, Frank A. J. L. *Michael Faraday: A Very Short Introduction.* New York, NY: Oxford University Press, 2010.

Jones, Jill. *Empires of Light*. New York, NY: Random House, 2003.

WEBSITES

Benjamin Franklin and Electricity
http://www.americaslibrary.gov/aa/franklinb/aa_franklinb_electric_1.html

The Library of Congress provides a brief history and explanation of Franklin's contributions to the study of electricity.

The Faraday Museum
http://www.rigb.org/visit-us/faraday-museum

The Faraday Museum's website is devoted to the life of Michael Faraday and the history of science and includes many online exhibits, videos, games, and public courses.

Technology and Medicine
http://www.sciencemuseum.org.uk/broughttolife/themes/technologies

An online exhibit on technological innovations in medicine, including the use of electricity, from the eighteenth through the twentieth centuries.

Thomas Edison National Historic Park
https://www.nps.gov/edis/index.htm

This site, maintained by the National Park Service, features online recordings, multimedia, essays, and other material related to Edison's life and times.

Bibliography

Bodanis, David. *Electric Universe*. New York, NY: Crown Publishers, 2005.

Camenzind, Hans. *Much Ado About Almost Nothing: Man's Encounter with the Electron*. Portland, ME: Booklocker.com, 2007.

Casarett, David. *Shocked: Adventures in Bringing the Recently Dead Back to Life*. New York, NY: Current, 2014.

Fara, Patricia. *An Entertainment for Angels: Electricity in the Enlightenment*. New York, NY: Columbia University Press, 2002.

Fellow, E. P. "Mark Cowley Lidwell: Inventor of the Pacemaker." Oct. 27, 2010. https://epfellow.wordpress.com/2010/10/27/mark-cowley-lidwill-inventor-of-the-cardiac-p/.

Freeberg, Ernest. *The Age of Edison: Electric Light and the Invention of Modern America*. New York, NY: Penguin Press, 2013.

Goleman, Daniel. "The Quiet Comeback of Electroshock Therapy." *New York Times*, August 2, 1990. http://www.nytimes.com/1990/08/02/us/health-the-quiet-comeback-of-electroshock-therapy.html?pagewanted=all.

Gribben, John. *The Scientists*. New York, NY: Random House, 2002.

Hirshfeld, Alan. *The Electric Life of Michael Faraday*. New York, NY: Walker and Co., 2006.

Isaacson, Walter, ed. *A Benjamin Franklin Reader*. New York, NY: Simon & Schuster, 2003. James Clerk Maxwell Foundation. "The Impact of Maxwell's Work." Retrieved November 16, 2016. http://www.clerkmaxwellfoundation.org/html/maxwell-s_impact_.html.

Jenkins, John D. *Where Discovery Sparks Imagination: A Pictorial History of Radio and Electricity*. Bellingham, WA: American Museum of Radio and Electricity, 2009.

Johnson, Jessica P. "Animal Electricity, Circa 1781: How an Italian Scientist Doing Frankenstein-like Experiments on Dead Frogs Discovered That the Body Is Powered by Electrical Impulses." *The Scientist*, Sept. 28, 2011. http://www.the-scientist.com/?articles.view/articleNo/31078/title/Animal-Electricity--circa-1781/.

Jones, Jill. *Empires of Light*. New York, NY: Random House, 2003.

King, Gilbert. "The Rise and Fall of Nikola Tesla and His Tower." *Smithsonian Magazine*, Feb. 4, 2013. http://www.smithsonianmag.com/history/the-rise-and-fall-of-nikola-tesla-and-his-tower-11074324/?no-ist.

Marquard, Brian. "Matt Nagle, 27: With Tenacity, Courage, He Broke Barriers." *Boston Globe*, July 26, 2007. http://archive.boston.com/news/globe/obituaries/articles/2007/07/26/matt_nagle_27_with_tenacity_courage_he_broke_barriers/.

NASA Godard Space Flight Center, Magnet Earth. "Oersted and Ampere Link Electricity and Magnetism." Retrieved November 16, 2016. http://www-istp.gsfc.nasa.gov/earthmag/oersted.htm.

Nickerson, Colin. "Everyday Nanotechnology." *Boston Globe*, Aug. 17, 2009. http://archive.boston.com/business/technology/articles/2009/08/17/nanotechnology_coming_soon_to_a_product_you_use/.

Nobel Foundation. "Nobel Prize in Medicine or Physiology, 1963." Retrieved November 16, 2016. https://www.nobelprize.org/nobel_prizes/medicine/laureates/1963/.

Shorter, Edward. *Shock Therapy: A History of Electroconvulsive Treatment in Mental Illness.* New Brunswick, NJ: Rutgers University Press, 2007.

Stanford Encyclopedia of Philosophy. "Herman von Helmholtz." Retrieved November 16, 2016. http://plato.stanford.edu/entries/hermann-helmholtz.

Stewart, Jeff. *$E=MC^2$: Simple Physics*. Pleasantville, NY: Readers Digest Books, 2010.

Unglow, Jenny. *Lunar Men: Five Friends Whose Curiosity Changed the World*. New York, NY: Farrar, Strauss, & Giroux, 2002.

Index

Page numbers in **boldface** are illustrations. Entries in **boldface** are glossary terms.

About the Author

Patrice Sherman is the author of several books for young readers on topics ranging from colonial America to mythological creatures. Her picture book, *Ben and the Emancipation Proclamation*, won the 2011 Once Upon a World Award from the Simon Wiesenthal Center in Los Angeles. In her former career as an archivist, she worked for a variety of universities and museums. She currently lives in Cambridge, Massachusetts, where she works as a freelance writer. You can find out more about her at http://www.patricesherman.com.